The Future of Diversity

The Future of Minority Studies

A timely series that represents the most innovative work being done in the broad field defined as "minority studies." Drawing on the intellectual and political vision of the Future of Minority Studies (FMS) Research Project, this book series will publish studies of the lives, experiences, and cultures of "minority" groups—broadly defined to include all those whose access to social and cultural institutions is limited primarily because of their social identities.

For more information about the Future of Minority Studies (FMS) International Research Project, visit www.fmsproject.cornell.edu

Series Editors:

Linda Martín Alcoff, Hunter College, CUNY
Michael Hames-García, University of Oregon
Satya P. Mohanty, Cornell University
Paula M. L. Moya, Stanford University
Tobin Siebers, University of Michigan

Identity Politics Reconsidered
edited by Linda Martín Alcoff, Michael Hames-García, Satya P. Mohanty, and Paula M. L. Moya

Ambiguity and Sexuality: A Theory of Sexual Identity
by William S. Wilkerson

Identity in Education
edited by Susan Sánchez-Casal and Amie A. Macdonald

Rethinking Chicana/o and Latina/o Popular Culture
by Daniel Enrique Pérez

The Future of Diversity: Academic Leaders Reflect on American Higher Education
edited by Daniel Little and Satya P. Mohanty

THE FUTURE OF DIVERSITY

ACADEMIC LEADERS REFLECT ON AMERICAN HIGHER EDUCATION

Edited by
Daniel Little
and
Satya P. Mohanty

THE FUTURE OF DIVERSITY
Copyright © Daniel Little and Satya P. Mohanty, 2010.
All rights reserved.

First published in 2010 by
PALGRAVE MACMILLAN®
in the United States—a division of St. Martin's Press LLC,
175 Fifth Avenue, New York, NY 10010.

Where this book is distributed in the UK, Europe and the rest of the world, this is by Palgrave Macmillan, a division of Macmillan Publishers Limited, registered in England, company number 785998, of Houndmills, Basingstoke, Hampshire RG21 6XS.

Palgrave Macmillan is the global academic imprint of the above companies and has companies and representatives throughout the world.

Palgrave® and Macmillan® are registered trademarks in the United States, the United Kingdom, Europe and other countries.

ISBN: 978-0-230-62068-1

Library of Congress Cataloging-in-Publication Data

    The future of diversity : academic leaders reflect on American higher education / edited by Daniel Little and Satya P. Mohanty.
      p. cm.—(Future of minority studies)
      ISBN 978-0-230-62068-1 (hardback)
     1. Education, Higher. 2. Minority college students. 3. Cultural pluralism. 4. Multicultural education. 5. Racism in higher education. I. Little, Daniel. II. Mohanty, Satya P. (Satya Prakash), 1954–

LB2326.4.F87 2010
378.1′982900973—dc22
                                                                  2009046870

A catalogue record of the book is available from the British Library.

Design by Newgen Imaging Systems (P) Ltd., Chennai, India.

First edition: June 2010

10 9 8 7 6 5 4 3 2 1

Printed in the United States of America.

Contents

Preface ... vii

Introduction: The Future of Diversity ... 1
Satya P. Mohanty

1. Universities and Democratic Culture ... 19
   Nancy Cantor

2. From Diverse Campuses to Integrated Campuses: How Can We Tell if We Are "Walking the Walk"? ... 41
   Jeffrey S. Lehman

3. Is Diversity without Social Justice Enough? ... 51
   Michael Hames-García

4. Equity and Excellence from Three Points of Reference ... 69
   Daniel Little

5. Prestige and Quality in American Colleges and Universities ... 83
   Steven J. Diner

6. Embracing the Commitment to Access, Diversity, and Equity in Higher Education ... 89
   Muriel A. Howard

7. Diversity and Excellence in American Higher Education ... 97
   Eugene M. Tobin

8. Educational Inequality and Three Ways to Address It ... 109
   Michael S. McPherson and Matthew A. Smith

9. Consuming Diversity in American Higher Education ... 123
   Gregory M. Anderson

10. College Access, Geography, and Diversity ... 147
    Teresa A. Sullivan

11	Higher Education and the Challenge of Inclusion *Marvin Krislov*	159
12	Notes from the Back of the Academic Bus *William A. Darity, Jr.*	173
13	Constructing Junior Faculty of Color as Strugglers: The Implications for Tenure and Promotion *Stephanie A. Fryberg*	181

Bibliography	219
Contributors	239
Index	245

Preface

This volume represents an effort at collaboration that goes beyond the customary boundaries of academic writing. The contributors include faculty and administrators; they represent a wide variety of colleges and universities; and they come from a wide range of disciplines as well. What they have in common is a commitment to thinking innovatively and practically about the challenge of making universities more fully embracing of the many forms of human diversity present in our society, and a determination to help institute the changes that are needed.

Several of the essays and many of the topics included here were first raised at a day-long conference at Cornell University in 2005, which was inspired in part by the publication of *Equity and Excellence in American Higher Education* by William G. Bowen, Martin A. Kurzweil, and Eugene M. Tobin (University of Virginia Press, 2005). The issues raised by Bowen and his coauthors are profound, and they demand thoughtful, pragmatic engagement. We need to create greater equity; and we need to go beyond that goal, toward greater inclusion and greater educational success in learning from manifold diversity.

Another main current underlying this volume is the ongoing work of the community of scholars from the many colleges and universities associated with the Future of Minority Studies (FMS) Research Project and the FMS Summer Institute (www.fmsproject.cornell.edu). FMS collaborators have made substantial progress in defining the challenges presented by the goal of creating a multicultural curriculum and university environment, and they have outlined some practical solutions to these challenges.

One of the most durable lessons learned by the experience of FMS concerns the value that comes from an extended and collaborative conversation transcending disciplines and institutions. All of the contributors to this volume agree that the issues raised here will not be resolved by a single moment of reflection and discussion. Rather, we need to learn from each other through extended dialogue, incorporating the insights of different institutional experiences and different forms of academic knowledge.

As the editors of this volume, we invite our readers to join our ongoing dialogue about the future of diversity. We ask you to share your thoughts with us about how these issues can be discussed on a national scale through the Internet or social media, and we will create an appropriate forum if sufficient interest is expressed. Please send your thoughts to fmsproject@cornell.edu and we will find appropriate mechanisms for sustaining the conversation.

We would like to extend a particularly sincere note of acknowledgment and appreciation to Harin Christine Song for her excellent editorial and research assistance in the process of assembling the volume. Her attention to detail and able assistance in the final stages of the volume are most appreciated.

Daniel Little
Satya P. Mohanty

Introduction: The Future of Diversity

Satya P. Mohanty*

In the early 1990s, two social psychologists conducted an experiment to see whether our society's negative racial stereotypes affect the learning experience of students in our educational institutions. They selected a group of black and white Stanford undergraduates and gave them a test made up of items from the advanced Graduate Record Examination in literature. The students had been statistically matched for ability, and since most of them were sophomores the GRE-based test was intentionally chosen so that it would be challenging and difficult for them. The psychologists—Claude Steele and Joshua Aronson—wanted to see whether there were differences in the way students of similar academic backgrounds but from different racial groups experienced a test that is supposed to be scientific and "objective." In particular, they wanted to see whether simple cues provided in the testing environment would be seen as innocuous or significant, and how these cues would affect the students' performance. The cues they provided casually were intended to refer indirectly to negative social images; their goal was to see, in short, whether negative social stereotypes were mere words, or if they had the power of sticks and stones. What they found was startling. When the test was given to the students as an abstract test of ability (that was the cue from the examiner), the black students in the group performed far less well than the white students. When, however, they presented the same test as a study of "how certain problems are

---

* **Satya P. Mohanty** has taught at Cornell since 1983, where he is currently Professor of English. He is one of the founding members of the Future of Minority Studies (FMS) Research Project (2000–) and has been the Director of the national FMS Summer Institute since 2005. His scholarly work deals in part with the relationship between minority identities and social justice.

generally solved," with a clear statement that the task did not measure intellectual ability in general, the black students' performance improved dramatically and now their scores matched those of the white students.[1]

Experiments like this one have been carefully replicated by researchers in various countries and they consistently produce the same measurable effect—not only in the case of racial stereotypes but also in those concerning gender and class. Psychologists call this phenomenon "stereotype threat." It impairs performance as long as it is included, even casually, in the setting in which learning and evaluation occur. A French psychologist later found the same effect of stereotype threat when he studied lower-class French college students and measured their verbal skills: their performance suffered when the threat was present and it improved and became normal when the threat was removed. Similar experiments have been carried out with women and men (in relation to domains such as science and mathematics). Interestingly enough, experiments reveal the power of such socially charged cues even when the group in question is not the target of a negative general stereotype. White students taking a math test with Asian American students performed poorly when they were told that that particular test was one in which Asians generally performed better. While there is no negative stereotype about white students' ability in math, the positive stereotype of Asian Americans did the trick, and—as in the case of the black students in the first study mentioned earlier—whites felt they were under the spotlight and the anxiety it produced made them perform worse than usual (unlike the white students who were not given the same cue about Asian Americans' performance on that test).

The series of experiments Steele and his colleagues conducted revealed to them that all our current beliefs about bolstering self-confidence and eliminating socially produced self-doubt are much less relevant to the learning context than we think. Instead, what the black students showed was that they were responding to their educational environment with "social mistrust." "When they felt trust," says Steele, summarizing the results of a series of these experiments, the students "performed well regardless of whether we had weakened their self-confidence beforehand. And when they didn't feel trust, no amount of bolstering of self-confidence helped" (52). He goes on to suggest that educational policy needs to recognize how "different kinds of students may require different pedagogies of improvement" (50), and that it should not be based on easy psychological generalizations about, say, the low self-esteem or self-confidence of some groups.

> Policies for helping black students [for instance] rest in significant part on assumptions about their psychology.... [T]hey are typically assumed to lack confidence, which spawns a policy of confidence building. This may be useful for students at the academic rearguard of the group. But the psychology of the academic vanguard appears different—underperformance appears to be rooted less in self-doubt than in social mistrust. (52)

Steele says that we need to think about "fostering racial trust" (52) if we want to improve the educational environment for vast numbers of American college students. This proposal—and the research on which it is based—goes to the heart of the discussion of "diversity," which is the focus of this volume. For social trust or mistrust are not merely attitudinal matters, to be left up to those who are affected by them, that is, the students; trust and mistrust—as we see in the case of the cues provided in the psychology experiments—are produced by our actions as teachers and administrators, and they reveal much more than our personal intentions as individuals. As many have argued in recent decades, trust is a social achievement and it takes us beyond our contractual obligations to be legally fair.[2] Trust and mistrust are often defining characteristics of the environment in which we all live and function, and they can exist even in the absence of overt discrimination. Social mistrust is often the net effect of a series of half-conscious acts. The presence of stereotypes alone is not enough; stereotype threat is the product of the social stereotypes *and* the ways they are reinforced by the partly unconscious cues we provide to students. In order to think about the "future of diversity," then, we need to look carefully at how our institutional attitudes and practices can be changed so that our colleges and universities can foster trust and effectively practice the "different pedagogies" that different "kinds of students" need. This is something every good teacher knows about the classroom, but when it is raised as a question about the college or the university campus as a whole, it makes us rethink the meaning of social diversity as a cultural ideal. Far from being content with recruiting greater numbers of socially underprivileged students, staff, and faculty, we need to see the ideal of social trust as a positive challenge to reimagine the culture of our campuses, to envision a culture that will be more conducive to learning precisely because it is more open, democratic, and genuinely attentive to the experiences of different social groups. Diversity and its future need to be rethought not only through the perspective of access (admissions, recruitment, financial aid, etc.) but also—and equally importantly—through the perspective of the campus as a learning environment for learners "of various kinds."

This volume contains essays by academic leaders from a variety of American institutions on both these perspectives—access and the culture of learning. How do we broaden access to more kinds of social groups? How do we make our campuses more genuinely inclusive? How do we conceive social diversity as a valuable educational resource, rather than a problem to be managed or solved? How, finally, do we replace the mistrust many feel—and the inequality of access, opportunity, and experience it points to—with the kind of social trust on which all learning, and indeed the very ideal of democracy, depends? These are big and general questions, and the prominent academics who have contributed to this volume—university and foundation presidents, deans, leading scholars—address them by drawing in part on their own specific experiences. They review what we have all learned from recent history—from the Supreme Court's verdict on the University of Michigan's use of affirmative action to experiments on various campuses involving students from different cultural backgrounds—and they make concrete proposals for the future.

Chapters 1, 2, and 3, by Nancy Cantor, Jeffrey Lehman, and Michael Hames-García, respectively, deal directly with the challenge of imagining a diverse campus as a valuable and unique learning environment, one that is in effect a social laboratory of sorts. Cantor—the President of Syracuse University and former Provost of the University of Michigan during the critical period when the recent Supreme Court cases were being prepared—cautions us against seeing diversity through "simplistic exercises in counting and balance" and argues that university campuses have a special role to play in building the future of our multicultural and diverse society. "In many cases," she argues, "college will be the first and best opportunity for young women and men (not to mention their faculty) to learn to affirm—rather than fear and privilege—difference, and to confront our common fates." Drawing on her own experiences at Michigan, Illinois (where she also served as Chancellor), and Syracuse, and on the most recent work in educational psychology, Cantor outlines general principles for building "healthy group dynamics"—an understanding of which, she argues, is critical "if we are to open up our institutions (and the power within them and conferred by them) and transcend the destructive fault lines of our society, thereby building the capacity for—and trust in—democratic culture beyond the campus." At the heart of her essay is the claim that the campus culture needs to be organized in such a way that it respects the "delicate balance between strong group identification and vibrant inter-group exchange." Like many psychologists, Cantor affirms the importance of group identification for the psychological well-being of those who are from socially marginalized groups, thus implicitly rejecting the popular notion that group identities are necessarily opposed to the nonparochial ethical perspective required of citizens of a democratic society. She also focuses on the importance of "normalizing" conflict, of raising— through "mutual respect and healthy interaction"—our consciousness of conflict so that we see it as a potential source of knowledge, a vitally important knowledge in a democratic society that thrives on difference (of background, of views, of life experiences).

Jeffrey Lehman—former President of Cornell as well as former Dean of the University of Michigan Law School during the *Grutter v. Bollinger* Supreme Court case—adds to Cantor's perspective on diversity by reminding us of Justice Scalia's skepticism (in his dissenting opinion in Grutter) that our campuses are indeed laboratories of a diverse culture and not in fact endorsers of "tribalism and racial segregation." Lehman says that taking Scalia's charges seriously should encourage proponents of diversity to demand a more rigorous self-evaluation of our efforts to produce an integrated culture of learning. Focusing on what he calls "integration pods" on campus— such as the campus cafeteria, where groups of students from different backgrounds often interact—he urges us to examine not just numbers but rather the way such pods affect the life of an individual student over time: "One must resist the psychological temptation to fixate on indicators of failure [of cross-racial integration]. One must not fixate on the homogeneous lunchroom tables.... Rather, one must try to see the entire picture, over time." Reflecting on his experiences at Cornell, especially of student-led efforts to forge an alliance of Muslim and Jewish students, he highlights the importance students attached to the supportive presence and participation of faculty. As long as they are not too "heavy-handed," he points out, "faculty members can have an impact on the culture of a campus. They can gently but effectively nudge their students in the direction of a daily ebb and flow. And they can nurture the integration pods they see on campus." Concluding with the assertion that campuses should be not merely diverse but also integrated—defined by "a flourishing, integrated, learning environment that is characterized by curiosity, civility, and a shared commitment to understand and appreciate the complex truths that define our world"—Lehman reminds us that beyond the challenge of the admission of a diverse student body lies the hard task of achieving a genuinely diverse culture, and to do that we need more empirical investigation to measure our successes and failures.

The author of chapter 3, Michael Hames-García, also focuses on the ideal campus culture, one that would consider social diversity a resource rather than a problem to be tackled, but he raises a basic challenge: "Is social diversity without social justice enough?" Examining his own experience as a faculty of color at a major research university, the University of Oregon, where he is Chair of the Ethnic Studies department, as well as his experiences as an undergraduate at a private liberal arts college and as a graduate student at an Ivy League school, Hames-García provides a trenchant critique of the current practice of separating the offices of "Diversity" from the main research mission of the university. His argument is that this makes Diversity Officers of most colleges and universities academically irrelevant and hence less effective. It may also foster a bureaucratic mindset that makes equity and diversity offices "get in the way of building substantive links between research faculty and multicultural student affairs." Hames-García makes at least two far-reaching proposals, one of which is easy to appreciate while the other—though tantalizingly bold—may be controversial in some quarters. His first proposal is quite simply that "[t]he research mission of the University needs to be front and center in multicultural affairs" and that "it is necessary for senior administrators to think of the positions as research positions in the hiring process." He sees the diversity offices as larger-scale versions of what Lehman would call "integration pods"—since he thinks of these offices as providing the links among various research units on campus that work on matters of race, gender, and social inequality. More radically, however, Hames-García goes on to argue that all diversity requirements in undergraduate curricula "should substantively address the nature of structural inequality, racism, power, and privilege, rather than emphasizing cultural diversity and tolerance of difference." His proposal contains the suggestions that (1) diversity offices be reconceived as offices with a mission to enhance social justice, and that (2) they—and the university in general—see student activism as a socially valuable resource and that they actively support and nurture it. Student activism, in other words, especially around identity issues, is less an example of the tribalism Justice Scalia deplores and more the kind of necessary group affirmation that Chancellor Cantor endorses. Hames-García's own experiences as a student "in a

Latina/o group, in a gay and lesbian student group, and a peace-and-justice residence hall enabled [him] to feel that even if there were very few students like [him] on campus, [he] at least had a place (or places) where [he] could feel supported and affirmed." Supporting socially marginalized students in their efforts to organize in these ways, he concludes, validates them in a pedagogically crucial sense—it makes them "more sophisticated activists and citizens."

Chapters 4, 5, and 6 focus on a subject that is not usually covered in many recent studies of diversity and higher education: the important role regional and nonflagship—rather than national and elite—institutions play in serving social needs and the ways they help us understand the varied nature of the "excellence" we seek in our educational contexts. Daniel Little, the Chancellor of the University of Michigan, Dearborn, talks about the particular role his institution plays in providing social mobility to a large group of (mostly minority and less affluent) Americans. He reflects on the role education can play in achieving social mobility, especially for first-generation college students, those who are the first in their families to go to college, and the special social role of nonflagship institutions:

> These institutions create a set of opportunities that mean that students from a range of backgrounds, from middle class to disadvantaged, can get a high quality undergraduate education for a total educational cost of about $8,000 per year, and can develop the preparation that will be needed for 'next steps' in professional schools, graduate schools, and working careers. UM-Dearborn is a source of genuine opportunity for the students we serve, and it provides them with a high-quality and effective education. (pp. 74–75, this volume)

Both Little and Steven Diner, the Chancellor of Rutgers University, Newark, focus on the unique functions of their own nonelite institutions, and they explicitly raise the question of value—of how we define educational excellence. Diner's campus serves first-generation immigrant families and provides an educational experience in which sociocultural diversity defines the learning environment, an environment that reflects the rich diversity of both American society in general and the increasingly globalized world in which we all live. But Diner points out that while his alumni recognize this environment and talk about it eloquently, the mainstream culture seems to lack the tools with which to measure its value. Thus, while *U.S. News and World Report*'s rankings system's focus on SAT scores skews it in favor of those institutions serving students from more privileged backgrounds, it lacks the ability to fully appreciate the value of diversity. It measures diversity, but it does not relate its diversity rankings to its quality rankings of schools. Diner points out, as does Little—and Muriel Howard, former President of another regional university, Buffalo State College—that their institutions are rich in pedagogical experimentation in part because their student body is so organically tied to the locations of the institutions. The faculty come to recognize this fact as an invaluable educational resource, as do faculty at most of our great urban institutions (the City University of New York is another good example), designing courses that explore the marvelous variety of immigrant and urban workers' experiences, memories, and cultural histories. Howard talks about how Buffalo State College consciously designed its mission and a comprehensive academic approach by focusing on its own history as a regional and urban institution; "through classroom discussions, curricular experiences, out-of-classroom activities, projects, discussions, or special mentor relationships," she writes, her colleagues "go beyond what is usually expected as a part of their regular job expectations or teaching assignments. The campus provides special financial incentives to students, faculty, and professional staff to support programs and projects that strengthen college diversity initiatives." The perspectives of the leaders of these nonflagship institutions suggest that conscious planning built around the recognition of the particular social role of their own institutions is a key ingredient of success. It also suggests, as Little puts it, that "excellence" in higher education may be the result of more things than the "inputs" that money and elite social status bring with them: students with high grades and SAT scores, faculty with the best educational credentials, and the best laboratories and libraries. Excellence, says Little, may have more to do with the ongoing project of mixing everything together in a certain way, envisioning the future as thoughtfully as we can. Like baking bread, success depends less on getting the most expensive ingredients and more on paying "constant attention to the process," which is the hard work put in by the leaders of various institutions. And here, elite social status provides no guarantee of success:

> [A]chieving a quality education is...like baking bread. The ingredients are the beginning. But constant attention to the processes is needed in order to keep the joint product working up to its maximum potential. If faculty lose focus on the importance of close intellectual relationships with their students; if they come to overvalue research time over classroom time; if deans and department chairs ignore signs of quality erosion; if faculty and leaders grow inattentive to important

developments in pedagogy, curriculum, and content; and if university leaders fail to consistently emphasize the priority of effective teaching and learning—then high-quality ingredients will still lead to mediocre bread. Put it another way: there is an important intangible aspect of educational quality that is measured by academic values and shared commitment to students' learning that is a feature both of the people of a successful university and its institutional makeup. And institutions differ greatly in this dimension! (pp. 76–77, this volume)

Chapters 4, 5, and 6 together clarify the nature of excellence in the educational setting. While elite status and financial resources are valuable, they do not guarantee a superior educational experience, for a quality education depends on a combination of factors, chief among which is the conscious planning and coordination by various levels of the campus leadership—in particular, the administration and the faculty. Moreover, these three essays in particular point to the crucial role played in any democratic society by regional and urban institutions in providing access and social mobility to immigrants and those from lower-income groups. If the goal is to reduce social inequality through education, then regional and urban universities need to be both recognized and supported by policy makers at not just the state level but also nationally.

The scandalous truth is of course that American national educational policy is weak precisely on a national level, since funding of public universities is increasingly being left entirely up to the states. What the recent economic downturn makes clear, however, is that American higher education, which has traditionally been the engine of the country's economic development, has fallen behind dramatically, and that is mainly because of the erosion of federal funding and our myopic social policies about lower-income groups. As the economist Paul Krugman points out in the *New York Times*, "[W]ith [the] weak social safety net [of the United States] and limited student aid, students are far more likely than their counterparts in, say, France to hold part-time jobs while still attending class."[3] Education and social mobility suffer due to a variety of related but largely invisible economic policy decisions, and the net effect is that American higher education is no longer available to the population at large. California's community colleges, for instance, have served for generations of lower-income families as a means of access to the state's admirable state university system; but now with the state's economic woes, transfer students are finding it impossible to enter the state universities. The phenomenon is a general one, with national effects, and it may leave its mark on this generation of students over their entire lifetime.

Krugman considers this predictable result of myopic national policy to be "a large gratuitous waste of human potential," and calls for Congress to take appropriate measures. "Education made America great," he points out, and goes on to issue a timely and urgent warning: "neglect of education can reverse the process."

The effects of poor educational policy on the lives of less affluent families is the direct or indirect focus of chapters 7, 8, and 9 in the volume. Eugene Tobin, former President of Hamilton College and currently an officer of the Andrew W. Mellon Foundation, draws on the book he coauthored with William Bowen and Martin Kurzweil (*Equity and Excellence in American Higher Education*[4]) to summarize their research findings about the positive effects of affirmative action on the racial composition of our campuses. But he also highlights the urgent need to pay attention to the issue of socioeconomic class as a measure of real diversity. Documenting the growing inequality of access to higher education in recent years, Tobin says that a big part of the problem is the difference in what is called "college preparation" between students from different class backgrounds.

> Young people whose parents' income is in the bottom quartile are half as likely to even take the SAT as those whose parents' income is in the top quartile. Our research (National Educational Longitudinal Study) indicates that the odds of taking the SAT and scoring over 1200—using the old scoring system with 1600 as the perfect score—are roughly *six times higher* for students from the top income quartile than for students from the bottom income quartile; and those odds are roughly *seven times higher* for students from the top income quartile than for students who are from the bottom income quartile *and* who are also the first in their families to attend college. (p. 100, this volume)

Noting the need to address social inequality in the broader national context, Tobin goes on to recommend, however, that at least the top universities, private and public, consider putting a "thumb on the admissions scale" by taking low-income status at least as seriously as we now take race. Research shows that students from less affluent backgrounds, once admitted, go on to do at least as well as those from more affluent ones. Broader considerations of social justice would necessitate that colleges and universities take class seriously in their definition of social diversity. Income-based preferences in admission, Tobin argues, should be seen as a necessary complement to the race-based programs that have been so successful in diversifying the major colleges and universities that have initiated such programs in recent decades.

The essay by Michael McPherson and Matthew Smith (chapter 8), both of the Spencer Foundation (McPherson, the President of the Foundation, is former President of Macalester College), surveys in great depth and detail the effects of socioeconomic deprivation on academic performance. Pointing out that in many cases some of these effects are long-term, since childhood poverty can produce physiological changes, they argue passionately for the thesis that the future of a better educational system lies beyond educational policy in the socioeconomic policies that address poverty itself. Merit, which is the chief criterion for admission used by most selective colleges and universities, is itself tied to socioeconomic status, and college preparation is—as Tobin points out as well—almost always determined by the economic resources of the student's parents. McPherson and Smith emphasize the need to design better admissions policies to improve access, but their most far-reaching point is that all socioeconomic policy is ultimately educational policy as well, especially given the growing gap between rich and poor in this country:

> For better or worse, the battle against educational inequality...devolves into a proxy battle against poverty.... It is unrealistic to expect educational inequality to disappear without massive attention to the underlying social and economic dynamics that fuel lower achievement for lower-income students. It is impossible to fully address educational inequality without addressing the inequalities from which it stems. (p. 120, this volume)

Writing from a similar perspective, Gregory Anderson argues (in chapter 9) for the need to distinguish the "fetishized consumption of diversity" in academic discourses from the recognition that race and class overlap to a significant degree in contemporary American society. Anderson, formerly a professor at Columbia University and an officer at the Ford Foundation and now Dean of Education at the University of Denver, issues a loud call for the recognition of the class basis of race in American society. Drawing attention to research (including his own past work) on such pervasive phenomena such as tracking, Anderson emphasizes the need to keep the class basis of race in mind when educational policy is designed. He calls for modes of assessment (such as the Posse Foundation's Dynamic Assessment Process) that do not rely exclusively on performance on standardized tests, and urges attention to factors that are "not easily gleaned from academic records" in selecting students for admission. He asks the elite, selective colleges and universities to directly confront the effects of defective educational methods, such as tracking and the growing socioeconomic inequality in the country, by devising more supple criteria for identifying need and talent, going beyond the obvious features available through the typical academic records.

> [T]he problem of restricted access to quality higher education opportunities for students of color from disadvantaged backgrounds requires bold, transformative leadership in the present. A refusal on the part of universities and colleges to take meaningful action at a critical juncture in American history in which millions of Americans have lost their jobs, their homes, and control over their lives, will have disastrous consequences especially, albeit not exclusively, for young people of color who constitute a growing proportion of the nation's poor. Although selective universities and colleges have been impacted negatively by the economic collapse, they nevertheless remain better equipped and resourced to address the challenges of access. It is therefore time to go beyond a rhetorical affirming of diversity to identifying innovative and practical ways to reduce the opportunity gap in higher education. (p. 141, this volume)

"Failing to do so," he warns, will create a limited form of diversity at elite institutions but will fail to produce real social change for the majority of Americans of color.

In her essay "College Access, Geography, and Diversity" (chapter 10), Teresa Sullivan, current Provost of the University of Michigan and the incoming President of the University of Virginia, proposes an innovative practical way to make admissions criteria more responsive to the diversity that exists in American society while remaining sensitive to the legal challenges faced by traditional affirmative action criteria. Drawing on the examples of Texas's Top Ten Percent law, which guarantees admission to students graduating in the top 10 percent of every high school to the state's public universities, and her own university's practice of considering applicants' residential districts as one criterion among others to shape its goal of achieving a diverse student body, Sullivan—a sociologist and demographer—argues that "micro-geography" can help us, within the terms of the current laws, to refine our admissions criteria, especially since residential areas are often relatively homogeneous in socioeconomic status and educational background. One advantage of this additional way of measuring diversity in admissions, writes Sullivan, is that "increasing the number of high schools represented within a freshman class represents an important means of strengthening a university's links to its publics." The essay by Marvin Krislov (chapter 11), President of Oberlin College (and previously at the University of Michigan), emphasizes this very need to strengthen the connection between universities and the general public they serve; Krislov's point is that while the 2003 Michigan

Supreme Court decisions (*Gratz* and *Grutter*) provide relatively good and flexible guidelines for colleges and universities to apply affirmative action criteria in seeking racial diversity, public education programs "coordinated among institutions" may be necessary to convey the goals and ideals of diversity that shape the practices of educational institutions. Sullivan points out that in the state of Texas the Top Ten Percent law has been so popular that legislative efforts to overturn the law have not succeeded. Krislov's and Sullivan's essays highlight the need for universities to make a concerted effort to engage with the general public in order to counter reactionary attacks from the enemies of social diversity. Krislov notes both the difficulties of this task and the urgent need to undertake a coordinated effort:

> One of the challenges to "fixing" the access problem in the United States is the enormous range of problems, across states, and across public and private entities. This not only creates structural challenges, but also communications challenges. Higher education may not appear to have a unified or coherent message regarding its own inclusionary programs. Given the mix of private and public institutions, the multiplicity of messages is to some extent inevitable. However, the higher education community can and should promote better public understanding of, and support for, inclusionary policies and programs. (p. 167, this volume)

Sullivan and Krislov raise an important question: are leaders of colleges and universities doing enough to educate the public about the educational virtues of diversity and can they be more successful if they coordinate and combine their efforts on both the state and the national levels?

The final two essays (chapters 12 and 13) in the volume combine personal reflections with practical analysis; the authors are both faculty of color who are prominent in their respective fields. William Darity is an economist who holds an endowed chair at Duke; Stephanie Fryberg, a pioneering younger psychologist and an Assistant Professor in Psychology and American Indian Studies, teaches at the University of Arizona. Darity writes about the example of the culture of American Economics departments and analyzes the reasons why there are so few African American faculty in that discipline. In a frank and hard-hitting essay, Darity raises the same question that many of the essays in the volume have asked: does a blind acceptance of "merit" (narrowly understood) as the primary criterion of inclusion lead in effect to exclusionary social practices? Darity reviews the state of his profession—from hiring and promotion practices to the selection of gatekeepers, namely, the journal editors—and suggests that the current practices are built on "the implicit (and occasionally explicit) belief in the inferiority of the black scholar." He argues that while "the rhetoric that would be used to justify the complete absence of black scholars from the editorial boards of the American Economic Association journals would have something to do with the notion that the economists invited to serve all were selected on the basis of 'merit,'" the reality is that it is another instance of "the fetishization" of merit to rationalize "discriminatory outcomes." He concludes with a direct challenge to the economics profession, which he calls a "backward discipline" among the social sciences, mainly because of its narrow assumptions: the fetishized view of merit "assumes that there would be no improvement in the quality of the editorial practices of the journals and ultimately their content if the composition of their editorial boards were different." He points out that "other disciplines in the social sciences have begun to reach a different conclusion" and goes on to list the number of key theoretical concepts that would be missing from the social sciences if the contributions of black scholars were to be left out: "discounted dynamic programming, stereotype threat, oppositionality, surplus labor, the dual economy, the plantation economy, color blind racism, programmed retardation, modernity, blaming the victim, educational subnormality, deficit models, resiliency, social capital, legacy effects, neocolonialism, postcolonial melancholia, double consciousness, preemptive extermination, racialization, and cultural representation."

Stephanie Fryberg writes from her perspective as a junior faculty of color about experiences that senior faculty and administrators need to understand if they are to create a climate where racial "mistrust"—the phenomenon we identified in the experiments of Claude Steele and his colleagues—can be removed. Fryberg does not refer to trust and mistrust explicitly, but she identifies stereotypical images of junior faculty of color that have had a major effect on her own career. She draws on her own expertise as a social psychologist to review the relevant literature in the social sciences that would help us understand such stereotypes and their effects, and concludes with practical proposals that will be useful for not only junior scholars of color who face similar challenges on their own campuses but also senior faculty and administrators who wish to make their institution more genuinely open and inclusive. Fryberg outlines the "professional and psychological toll" that faculty of color pay for the negative cultures that exist on their campuses, and draws attention to the things that can be done to survive as well as to transform them. At the core of the problem, Fryberg argues, are narrow and false assumptions about what merit is, assumptions that blind those in positions of power to the complex social beings that their junior faculty of color are:

> Universities must reexamine the pervasive ideas about academic excellence and diversity efforts. First, universities need to identify the objectives it hopes to achieve through diversity efforts. Is the goal to increase the representation on campus or to foster an academic climate that benefits the entire campus community? Second, what is the current state of diversity efforts with respect to the mission of the university? Do diversity and excellence go hand-in-hand or do ideas about excellence appear to be in conflict with ideas regarding increasing diversity? Third, what are the pervasive ideas about success? Do images of "successful academics" encourage expansive forms of knowledge generation, innovative pedagogical approaches, and new pathways for mentorship and success, or do they advocate traditional and often ineffective (or outdated?) models of success? Fourth, what language is used to discuss those individuals historically underrepresented at the university? (p. 208, this volume)

Our efforts to achieve the social diversity on our campuses cannot be successful if we do not question our deeper assumptions about what success is and what produces an effective culture for the work of scholarship and teaching. Such work is not done by abstract individuals but by socially embodied beings, with socially produced strengths and vulnerabilities, and any attempt to think about the culture of a campus must focus on the actual experiences of such faculty—and students. This requires a rethinking of some of our most basic theoretical assumptions as well as a reexamination of our traditional habits and practices.

One of these theoretical assumptions concerns the nature and value of what is called "objectivity." It is possible to worry that while taking the subjective experiences of students and faculty of color into account may improve the campus culture in some respects, it compromises the objectivity of our approach as senior faculty or administrators. That worry is based on the understanding of objectivity as pure "neutrality," and there are reasons to doubt that this conflation of objectivity with neutrality is intellectually justified. Modern philosophers writing about objectivity concur with those historians and social scientists who talk about the need to see objectivity as a context-sensitive value rather than as the product of an abstract and a-contextual attitude of neutrality.[5] So in contexts where unfairness is built into the environment because of half-conscious habits and practices—which echo and reinforce prejudices and stereotypes prevalent in the social mainstream—genuine objectivity may itself be the product of conscious effort to examine our assumptions rather than of a neutral approach—as evidenced, for instance, in "color blind" or "gender blind" policies. What seems fair and just to a member of one social group is not in fact experienced in the same way by members of a group that is, say, the target of negative social stereotypes—as we saw in the psychological experiments of Claude Steele and his colleagues. It may be useful, then, to return to that series of experiments.

At the end of the series of tests that revealed the damaging effects of stereotype threat on black students, the researchers wanted to see how to reduce stereotype threat in the testing environment. They found something very interesting: the black students who felt racial mistrust did not respond favorably to gratuitous or conventional positive feedback ("there are many nice things about your essay," e.g. as a cushion for the negative comments that follow). But they responded very favorably to explicit statements about what the standards of evaluation were, why and how these standards were high, and how the student could improve his or her work and achieve these standards (i.e. an explicit statement that the student will not be viewed through racial lenses). It is important to note that such a clarification was unnecessary for the white students, but it seemed to make all the difference to the black students—and this assurance that they would not be evaluated stereotypically made the black students much more motivated and eager to learn and improve, even more than the white students. The experimenters, instead of pretending that the evaluative context was unbiased and neutral, worked hard to address the socially produced deficit of trust; in doing this, they prepared the conditions for greater objectivity in their testing. Trust was made possible by consciously *and sincerely* making their effort and their standards explicit, and the students reacted as genuine learners. In fact, they reacted with greater enthusiasm for learning than did their white peers, which reveals, according to Steele, just how profoundly socially prevalent ideas affect even the carefully circumscribed learning environment on our campuses, as well as how reversing these effects requires a deeper commitment to objectively high standards and a genuine fairness (rather than a blind neutrality).

> High standards, at least in a relative sense, should be an inherent part of teaching, and critical feedback should be given in the belief that the recipient can reach those standards. These things go without saying for many students. But they have to be made explicit for students under stereotype threat. The good news of this study is that when they *are* made explicit, the students trust and respond to criticism.

Black students who got this kind of feedback saw it as unbiased and were motivated to take their essays home and work on them even though this was not a class for credit. They were more motivated than any other group of students in the study—as if this combination of high standards and assurance was like water on parched land, a much needed but seldom received balm. (53)

In thinking about the culture of a genuinely inclusive learning environment, then, the first great challenge for us may be to remind ourselves that what is needed is not so much sentimental partiality as—ultimately—greater objectivity; an assurance of genuine fairness restores social trust. The future of diversity on our campuses depends on our thinking hard about restoring to education and learning the healthy environment of mutual trust and respect in which alone they can thrive. And while social forces beyond our immediate control do much to diminish this trust, the joy—indeed the magic and mystery—of learning is that it can transcend such forces. The world pervades our classrooms and our laboratories, but it does not wholly determine what can be achieved in them.

The essays in this volume are a testimony to this faith in the power of education to transcend the social forces that limit us as individuals. They contain new practical proposals even as they raise questions of a more theoretical nature. But more than anything else these academic leaders suggest the need for all of us—including readers of this book—to coordinate our efforts to reimagine our campuses and to work toward making them the laboratories that they can be—of the future society we hope to build. Social diversity, they suggest, is not a "problem" to be solved; it is an enormous resource that is waiting to be tapped. From admissions to sports to the designing of the curriculum and of noncurricular interactions, the practical challenges posed by what we call "diversity" are the gateways to a more democratic future.

Notes

1. Claude M. Steele, "Thin Ice," *The Atlantic Monthly* (August 1999): 44–54.
2. The best discussion is by the philosopher Annette Baier; see esp. "Trust and Anti-Trust," *Ethics* 96 (January 1986): 231–60. See also Francis Fukuyama, *Trust* (New York: Free Press, 1995) and Danielle Allen, *Talking to Strangers: Anxieties of Citizenship since Brown v Board of Education* (Chicago, IL: University of Chicago Press, 2006).
3. "The Uneducated American," Op-Ed Column, *The New York Times*, October 9, 2009.

4. William Bowen, Martin Kurzweil, and Eugene Tobin, *Equity and Excellence in American Higher Education* (Charlottesville, VA: University of Virginia Press, 2005).
5. An excellent survey of some of these issues can be found in Louise Antony, "Quine as Feminist: The Radical Import of Naturalized Epistemology," in *A Mind of One's Own*, ed. Antony and Charlotte Witt (Boulder, CO: Westview, 1993). For a briefer account, see my "Can Our Values Be Objective?" in *Aesthetics in a Multicultural Age*, ed. Emory Elliott, Lou Caton, and Jeffrey Rhyne (New York: Oxford University Press, 2001).

1

UNIVERSITIES AND DEMOCRATIC CULTURE

*Nancy Cantor**

We live in a time of paradoxes and contrasts that have considerable consequences for the health of our democratic community. Although the demographic landscape of our nation has become less and less white, our patterns of residential segregation have not changed. According to data gathered by the Harvard Civil Rights Project,[1] we have created a growing number of "apartheid public schools,"[2] where isolation and poverty have joined forces to close off the traditional route to upward mobility in our society, a tragedy compounded by the increasing value of education in a "knowledge economy."

Around the world, violent ethnic and religious intercultural conflicts plague the lives of millions at the very time we find ourselves facing ever more resistance at home to the peaceful integration of Islamic and other non-Western traditions. We know the importance of intellectual diversity as a foundation for democracy, but we are losing our willingness to work through difference in civil dialogue. The creation of diversity is too often reduced to simplistic exercises in counting and balance.

In this context, higher education (private as well as public) has a critical—one might say, urgent—role to play as a public good. In many cases, college will be the *first* and *best* opportunity for young

---

*****Nancy Cantor,** Chancellor of Syracuse University and former Chancellor of the University of Illinois at Urbana-Champaign, has worked to forge new understandings of the university as a public good, promoting community engagement, racial justice and diversity, the status of women, the creative campus, and sustainability. As Provost of the University of Michigan, she was closely involved in the university's defense of affirmative action before the Supreme Court.

women and men (not to mention their faculty) to learn to affirm—rather than fear and privilege—difference, and to confront our common fates.

UNIVERSITIES AND BUILDING DEMOCRATIC CULTURE

If we are to shape and keep a democratic culture, we must nurture understanding and create support for the principles of freedom and justice that are embedded in our nation's founding documents. The task starts on campus and must spread beyond. A hallmark of democratic culture is that difference can be respected and tolerated.[3] Therefore, it is essential that we not only comprehend and practice the individual rights to freedom of speech, expression, and association, but that we also take on the responsibilities to assure those benefits of participation to all individuals, as individuals and as members of groups. Universities have a special role to play in educating about difference and facilitating all kinds of vigorous *intergroup*, as well as individual, exchanges.[4]

If universities are going to build democratic cultures that make the most of newly achieved access to opportunity for diverse students, faculty, and staff, we will have to understand better than we do now how to embrace the principles of healthy group dynamics. An understanding of these principles is also critical to society if we are to open up our institutions (and the power within them and conferred by them) and transcend the destructive fault lines of our society, thereby building capacity for—and trust in—a democratic culture beyond the campus.[5]

BALANCING INTRA-GROUP AND INTERGROUP COMMITMENTS

Before considering some specific programmatic ways to build democratic culture on campus, I want to underscore one of my foundational assumptions. Democratic culture is built on the constructive coexistence (and absence of contradiction) between the affirmation of many groups and the necessity for healthy intergroup relations. A strong democratic culture thrives on the strength and resilience of its many cultures and groups. And, as social psychologists have long noted, individuals also derive considerable personal well-being and strength from the feelings of social connection, belongingness, and commitments to others with shared values and life experiences that characterize group life.⁶ Actively taking part in the valued tasks of one's group is a good predictor of health, productivity, and happiness.⁷

Therefore, I would posit that democratic culture hinges on the opportunities it can provide for individuals to identify with and derive affirmation from their own group commitments. By the same token, democracy also strongly depends on the freedom of all to explore and cross boundaries of every kind: intellectual, social, and cultural.⁸ When groups become insular and serve as barriers to exploration, they thwart individual and societal growth and resilience. Democratic culture thrives on openness, and there is nothing more stultifying for individuals or institutions than insularity. At the same time, openness can lead to conflict; and a healthy democracy must tolerate and manage difference, not just sweep it under the rug. If we are to reap the real benefits of diversity, we must pay attention and lend our skills to the continual task of working through conflicts.

Empathy and Conflict

At the heart, then, of democratic culture and individual well-being is this delicate balance between strong group identification and vibrant intergroup exchange. And it is a delicate balance indeed. Psychologically, what makes our group identities so self-fulfilling—that is, the reinforcing nature of our shared values, opinions, experiences, and aesthetics—is often what makes us unappreciative of other groups and resistant to seeing things from their perspectives. Without even realizing it we become defensive, closing off or undermining constructive intergroup relations.

To cite a brief example, during my tenure as Chancellor of the University of Illinois at Urbana-Champaign, I was witness to the consequences of a long-simmering intergroup confrontation. It was a painful dispute that could and does happen, in one way or another, on every campus. The controversy at Illinois revolved around its "Indian" mascot/symbol, Chief Illiniwek, who brought tears to the eyes of thousands of students, alumni, and friends as he danced at half-time, and tears to the eyes of countless other students, faculty, and friends who felt devalued and offended by the same dance/symbol. The debate surrounding the Chief was fierce, seemingly endless, and sometimes brutally off-putting. Nothing seemed to work in this debate. Proponents of the Chief saw no reason to distinguish their experience of the Chief—as uplifting, and exciting, and honoring something—from the experiences of those who perceived the

Chief differently. And, more to the point, because the Chief, in their eyes, was not *intended* to be hurtful, they felt no need to analyze the impact it had on others.

What was that impact? It varied certainly from person to person, and proponents and opponents each pointed to their definitive source. Notwithstanding, here is what one member of the American Indian Center in Chicago had to say when he came to campus: "We are not mascots. We are people. There's a hurt there that people don't feel or see. They're falsifying our traditions. We don't do this stuff for entertainment."

It is fundamental to healthy intergroup relations (and therefore to democratic culture) that we understand the gap between (honorable) intentions and (negative) impacts. Just as important, we must each take responsibility for the *possibility* that others have experiences different from the ones we cherish. Supporters of the Chief, not evil people, were not unconventional, nor did they hold views that differed from many of their opponents on other issues. In fact, there was nothing more perplexing and commonplace in Illinois than the compartmentalization of personal attitudes about the Chief. Time and time again, I met staunch supporters of the Chief who saw themselves as, and even demonstrated themselves to be, staunch activists on diversity. It just so happened that the supporters of the Chief grew up loving something that was very hurtful to some others, and they did not see why. Unfortunately, however, there is no one more righteous (on any and all sides) than a person with *good intentions* encouraged by like-minded others and so this debate raged on, seemingly inoculated from any other ostensibly relevant questions of diversity and intergroup relations. Ultimately, the matter was resolved from the outside. Pressure from the National Collegiate Athletic Association led to the "retirement" of the Chief.

It is precisely the ubiquity of this experience that makes it so important. We must realize that even our best intentions may cause someone else's worst nightmares. In the process, as good as we are at privileging our own intentions (over their impacts on others), we often fail miserably to acknowledge the "good intentions" of others when there are hurtful impacts on us. Although we expect the world to forgive our well-intentioned "mistakes," we are rather unforgiving of others, especially those we fear or do not know. These patterns of intergroup relations make it difficult to build democratic cultures in diverse settings. They constitute a "natural" set of defensive barriers that education must neutralize. How then do we learn to lower the barriers, to decenter from our own intentions, and to gain empathy for "others"?

Understanding Difference

Our society remains largely segregated by race, ethnicity, religion, culture, and class in our K-12 schools, our neighborhoods, and our social spaces. When we do not live together, we do not know each other. As a consequence, many of us tend not to talk about difference. Even in academic courses that focus on these issues, both the time and structured format for sharing and learning are limited. Nonetheless, as Patricia Gurin has argued, the understandable approach of trying to mute intergroup differences and build overarching common identities across boundaries is neither effective nor desirable.[9] While the aim must still be to encourage perceptions of *common fate* and *interdependence* across groups, the path toward healthy intergroup relations cannot be built by denying difference.

Students, especially with minority group social identities, want and need to learn how to narrate their stories and experiences and to talk about their lived knowledge, struggles, and resistance in ways that will be heard and richly and respectfully understood.[10] To do this, and to lower the defensive barriers to intergroup exchange, they and all other members of the democratic cultures in which we hope to live must feel affirmed in their own groups. It is hard to trust others without a secure "place" to which to return periodically.

The standard reasoning behind muting difference is that conflict is to be avoided at all costs, and focusing on difference tends to bring forth tension and conflict. But there must be another way, and the intergroup dialogue curriculum that Gurin and her University of Michigan colleagues have pioneered is built on the assumption that intra-group affirmation and intergroup appreciation can go hand in hand. As I discuss in a bit more detail later, the intergroup dialogue addresses group differences not in debate, with winners and losers, but in a conversation that is revelatory for everyone involved.

The intergroup dialogue process brings us back to the intertwining of empathy and conflict, and to finding ways to tolerate both. For example, one of the thorniest aspects of intergroup dialogue is getting people to acknowledge that the world is not a level playing field—in fact, that it is un-level on many dimensions. Those discussions are essential to building genuine empathy of mind for (out)groups, but they raise defensiveness from (in)groups. The answer is not to pretend that we live in a group-less, color-blind, or equitable society, but rather to find a way for everyone to express their resentments while acknowledging that others also have vulnerabilities. The basis for conflict, if handled well, can become the path toward empathy.

Regardless of how hard this is to accomplish in a world as divided as ours, I do not see a path toward empathy that does not air conflicts, so all can learn about perceptions and experiences that otherwise remain hidden and/or defended against. Opportunities to build these "difficult dialogues"[11] across the divisive fault lines of our world are very rare, but universities can create them. In so doing, we can and should take a lesson from the cultural expressions that come into our daily lives through the arts—through movies, photography, theater, dance, and all kinds of music.

A powerful example of the ability of the arts to create new dialogues and draw us in, across history and in the most unlikely places, can be seen in a *New York Times* article by Jodi Wilgoren that described a production of "King Lear" at the Racine Correctional Institution in Wisconsin.[12] The 17 actor-inmates, who were doing time for kidnapping, homicide, drug dealing, and other crimes, called themselves the Muddy Flower Theater Troup, acknowledging that beautiful things can grow in unlikely places. One of the inmates, who played Lear's counselor Kent, expressed it this way: "There are no walls now. I'm in medieval England." Although the prisoners could not use swords and performed in a large room—not really a stage—they found their experience to be transforming. Some found themselves crying in front of others for the first time in their lives. In the prison yard they started calling each other Cornwall and Oswald and Goneril.

Universities are ideal places for the arts to serve as the medium, not just the reflection, of intergroup dialogue. Art offers an escape from the silencing that tends to come in "normal" society, making it possible to face highly charged and even taboo subjects. And everyone has some "standing" in the "conversation" that ensues. Somehow, we can confront even the most searing of intergroup and intercultural experiences with relatively little group defensiveness and more honesty in the context of the literary, visual, and performing arts. We accord more (human) standing to our "enemies" in the arts, exhibiting far more empathy of mind than is usually the case. We see the universality of vulnerabilities and of differences more clearly through the artistic lens and in so doing manage to affirm multiple groups and traditions.

Khalid Hosseini's best-selling novel *The Kite Runner*, chosen as the centerpiece of the Syracuse University Shared Reading Program for all incoming students in 2006, is an outstanding example of the power of the storyteller to astonish us by putting his readers—and his own characters—in the shoes of others, even those they think they know. This novel was subsequently used by CNY Reads, the largest "one book, one community" program in New York state, as its 2007 selection in programs for adult and high school readers. The novel tells the story of Amir, the son of a wealthy Pashtun businessman in modern Afghanistan, and Hassan, the son of his family's Hazara servant. Both boys have lost their mothers at birth and they play together almost like brothers, although Amir goes to school and Hassan does not. Amir takes the social distance between them for granted—until he opens one of his mother's old books and is "stunned to find an entire chapter of Hazara history."[13]

> In it, I read that my people, the Pashtuns, had persecuted and oppressed the Hazaras. It said the Hazaras had tried to rise against the Pashtuns in the nineteenth century, but the Pashtuns had "quelled them with unspeakable violence." The book said that my people had killed the Hazaras, driven them from their lands, burned their homes, and sold their women. The book said that part of the reason Pashtuns had oppressed the Hazaras was that Pashtuns were Sunni Muslims, while Hazaras were Shi'a. The book said a lot of things I didn't know, things my teachers hadn't mentioned. It also said some things I did know, like that people called Hazaras *mice-eating, flat-nosed, load-carrying donkeys*. I had heard some of the kids in the neighborhood yell those names to Hassan.[14]

The following week, when Amir shows the chapter to his teacher, "he skimmed through a couple of pages, snickered, handed the book back. 'That's the one thing Shi'a people do well,' he said, picking up his papers, 'passing themselves as martyrs.' He wrinkled his nose when he said the word Shi'a, like it was some kind of disease."[15] Although Amir remembers this moment later in life, his new knowledge of history produces no change in his treatment of his friend Hassan. *The Kite Runner* shows, in ways far more intimate than most journalistic accounts, how the ethos of a dominant culture can combine with ordinary emotions such as jealousy, fear, or love to paralyze some people from intervening to stop acts of unspeakable cruelty, even when these people know or even care deeply about the victims.

Moreover, there are, as is often the case in our volatile world, multiple layers of intergroup hostilities in contemporary Afghanistan—with both the (oppressed) Hazara and the (dominant) Pashtun sharing a fierce enmity for the marauding Taliban. What *Kite Runner*, therefore, manages to do in the most pressing of ways is to complicate greatly the American's monolithic post–9/11 view of all Afghanis as members of an (out)group of potential terrorists. The novel allows us to enter the minds of people living in and battling each other in a world unknown to many of us, except in the most simplified and stereotyped of ways. Hosseini also give us an empathetic view of the lives of Afghan refugees in our own country, post–9/11, as they struggle not only with the destruction of their homeland and life as they knew it, but with our (homogenizing) out-group enmity toward them.

It is hard to imagine a clearer, more penetrating portrayal of the complexity of intergroup hostility and the difficulties that all peoples have in appreciating difference, finding common ground, and acknowledging enmities that divide and destroy peace for everyone. The novel is full of conflict living side by side with empathy, and the experience of reading it forces one to consider how to stretch empathy to reach across the blinding barriers of intergroup defenses. How can we address this complexity, these fault lines, on our campuses? How can we develop the empathetic skills of our students and university community more broadly?

Building Democratic Cultures: Intergroup Dialogues and Mentoring

There are two arenas of university life, intergroup dialogue curriculum and faculty mentoring programs, that I believe serve as good illustrations of opportunities on campuses to build democratic culture. In each arena, there are opportunities simultaneously to affirm groups and to encourage intergroup exchange in ways that build trust. When this happens, the institution becomes more open and transparent, and participants, including those previously underrepresented in the mainstream of university life, see opportunities for sharing power and leadership, thus spreading more inclusively the sense of ownership of the institution.

Intergroup Dialogue Curriculum

Not surprisingly, as universities become more diverse, pervasive intergroup fault lines in our society are reflected more and more on campuses, leading many people to wonder about the value of diversity if students self-segregate by class, ethnicity, sexuality, religion, and culture, anyway. Some even argue that the intergroup conflicts—ever present just below the surface in our campus (and noncampus) communities—are harmfully exacerbated by emphasizing diversity.[16] No one should be surprised to find these fault lines replicated, nor should we blame students if they want to hang out with others just like them.

After all, many students come to campus with very little experience of intergroup interaction. In this country, their neighborhoods and their schools tend to be clustered by religion, culture, class, race, and ethnicity.[17]

Students from all kinds of backgrounds—majority and minority alike—may feel some degree of vulnerability around others not just like them. And these perceived vulnerabilities and fears of difference are often hidden under expressions of intergroup conflict, tension, and—occasionally—overt hostility. As Claude Steele and his colleagues have elegantly demonstrated, these intergroup dynamics are not only disruptive to the atmosphere on campus, they also constitute barriers to student achievement.[18]

As difficult as it may be to penetrate these intergroup dynamics, there is little that is more important to accomplish on our campuses today. Our mandate is to reap the full educational benefits of diversity and adequately prepare all students for living and working in a multicultural world, and to ensure that our minority students persevere. To do this, we must do on campus what we rarely do outside—that is, build a democratic culture of healthy interaction within and across the many groups of our increasingly diverse communities. If we can do this, then, as Patricia Gurin frequently notes, diversity becomes an educational resource embedded throughout the institution, like books in a library or faculty of quality, benefiting not only students' intellectual growth, but also better preparing them for citizenship.[19]

However, we cannot hope to do this by ignoring the societal fault lines we find on our campuses. We must tackle them with at least as much effort as we routinely put into teaching theoretical physics or music composition. Building a healthy democratic campus culture is no less labor intensive, but also no less rewarding. Gurin and her colleagues at Michigan, who have pioneered an intergroup relations curriculum that paves the way for trust between groups,[20] are leading a 10-institution evaluation of its effectiveness in building democratic culture—a project in which Syracuse is proud to be a participant.

The approach adopts an explicitly intergroup focus by bringing together members of groups that rarely interact in meaningful ways or may even be at odds. It acknowledges that these groups often occupy different positions in society, thereby uncovering patterns of structural inequality or cultural dominance that remain but often still go unnoticed. It also affords balance in representation of each group in the dialogue, so as to provide intra-group security and comfort as intergroup conflicts and differences are aired.

With this structure as the basic model, trained facilitators, who also benefit from the experience, can guide a *dialogue*—not a *debate*—in which participants are encouraged to explore similarities, differences, and conflicts within and between groups. The dialogues are supplemented with readings and reflections on structural inequalities and intercultural conflicts, both historical and contemporary. The goal is not to lay blame but to build empathy and understanding, including appreciation for vulnerabilities in the "dominant" group that frequently go unseen.

Students of color and others from "marginalized or targeted" groups often bring to the dialogue a serious interest and commitment to creating broader change, within their communities as well as across communities. One applicant for this program at Syracuse University, a junior, wrote:

> As a music education major and a LGBT studies minor I feel that the intergroup dialogue experience will give me the opportunity to really put to good use the pertinent everyday topics we have talked about in so many of my classes. Being able to openly engage in discussion about issues that are commonly neglected will not only help me become a better educator of young people in the future but also make me a better member of an ever changing society. As a member of the executive board of Pride Union it is essential that I be able to communicate to others about issues the LGBT community deals with. The intergroup dialogue would provide me with the skills needed to be a better and more informed leader within the LGBT community, as well as in the overall SU community.

The goal of this form of dialogic thinking and interaction is to "normalize" conflict as a part of life in a diverse society and that can be managed with mutual respect and healthy interaction. Accordingly, the balancing of group representation is not an end in itself, intended to give each group an equal shot at persuasion, as seems to be the objective in recent public calls for an "Academic Bill of Rights" on campuses.[21] Instead, by affirming groups both individually and in relation to others, and seeing how much people differ even within a group, some of the monolithic thinking about groups breaks down and the ground is laid for building common cause.

Preliminary results of the Multiversity Intergroup Dialogue Project show that all students benefit from talking across difference in these groups, regardless of the histories and statuses they bring, and are able to apply these insights and skills in other settings, including in other classes but also with family and friends as well. Participants said afterward that they think harder about how their own group identities affect their lives. They are also better able to discern the structural factors underlying group-based inequalities, such as those associated with race, ethnicity, gender, or sexuality, enabling students—including students from marginalized groups—to recognize their own privilege. Participation in these groups also promoted a far greater increase in empathy than was found in (control) groups of students who expressed the same interest in participating in the dialogue courses but were not chosen in the random-selection process.[22]

At Syracuse, in addition to the structured course curriculum on race/ethnicity, sexuality, and gender as part of the Michigan consortium, we are pursuing these intergroup dialogues in residence halls, in collaborations in local schools with the Inter-Faith Works of Central New York's Community-Wide Dialogues to End Racism,[23] through the work of our Program on Religion and Society in creating "difficult dialogues on religious pluralism," and in the context of our student-led Team Against Bias (that arose some time ago after several bias incidents involving racism and homophobia), and we hope to extend support for dialogues on diversity each year for the next several years.[24] There is much work to be done in building a democratic culture on campus and with our community through dialogue.

While "numbers" are good indicators of access, institutions of higher education must take additional steps to ensure full participation and to experience the full benefits of diversity. As Patricia Gurin and her colleagues in the Multiversity Project suggest, intergroup dialogue projects are excellent avenues for this next critical step. Whether we are talking about the confidence to tell one's story or the realization of privilege, experiences in these groups can empower their participants to embrace perspectives larger than themselves, rooted in empathy and the common interests of the group. These are essential for any democracy and for any just society. Although students usually begin with a strong sense of their rights as individuals, it takes collective effort and will to know how to exercise these rights responsibly so that others may have the same freedoms, how to engage those with privilege, and how to empower those without it. We need to be aware that the tables can and do turn as we negotiate social identities in a pluralistic, but largely divided, society.

Faculty Mentoring Programs

As we consider the landscape of intergroup relations and tensions for students on campuses today, it is also critical that we tackle similar issues for the faculty. Though the faculty and other leaders of our universities likely have more experience with diversity, they are certainly not immune to difficulties with the same fault lines that divide students. Frequently, though not always, differences fall out according to seniority, both professional and in terms of historical representation in the discipline. Junior faculty, faculty of color, and women in underrepresented fields feel relatively less secure, less embedded in the mainstream of departmental and university life. And just as this leads to group affiliations and some intergroup tensions for students, so too may it emerge in the faculty, though perhaps "disguised" as differences of opinion, lifestyle, or scholarly preferences.

To make this concrete, consider the number of "debates" that we have all witnessed in which senior faculty (often quite distinguished) bemoan how junior faculty simply do not appreciate excellence anymore. On the other side, many junior faculty feel that their senior colleagues do not fully appreciate (or perhaps fairly value) scholarship that crosses the line to practical engagement or that crosses disciplinary boundaries. These conversations sometimes unfold in the debate rather than dialogue mode—with one side upholding standards and the other side perceiving the institution as unresponsive at best and run by dinosaurs at worst. These debates in turn impede the integration of our newer faculty—new by seniority or access—into the mainstream of departmental and university life. And, the retention and development of these scholars is critical for the health of their disciplines and fields of study.[25]

One solution could be changing the way we mentor junior faculty and staff. Traditionally, we emphasize the transmission of information within the established hierarchy of departmental life: a senior faculty member tells a junior colleague how best to run his or her career. This faculty member may also have some considerable role in evaluating that junior colleague's record. From an instrumental perspective, this makes a great deal of sense. If the senior faculty member is conscientious, he or she can impart some very helpful "wisdom" about how things run and work in the institution—such as which committees to serve on and which to avoid, or how to get institutional research funds or teaching assistants.

Improved mentoring, however, will not be sufficient to reach the goal of opening up the system to new scrutiny from and engagement with junior colleagues who may have alternative ways to lead a career and/or do scholarship.[26] For example, departments are unlikely to figure out ways to support junior faculty doing interdisciplinary work or scholarship in collaboration with practitioners if the "received wisdom" of senior faculty prevails: those junior faculty who become "overly" involved in complicated interdisciplinary or activist scholarship jeopardize their quest for tenure. It is extremely hard for any of us to appreciate that the way we did it might not work for others.

In a one-on-one asymmetrical mentoring relationship, it is too easy for the powerful voice of the senior faculty member to feel challenged, for the junior faculty member to feel vulnerable, and for both sides to feel underappreciated—a recipe for intergroup conflict. When this situation is matched with a role for the senior person in the evaluation of the junior person, it is bound to exacerbate tensions and resentments from both parties. The senior mentor will then be prone to resolve the tension by devaluing the junior person—perhaps even generalizing to this "new generation"—and the junior person will not only feel unwelcome in the department, but he or she will likely perceive a no-win situation that may be immobilizing.

However, these pitfalls might well be overcome if we were to acknowledge structural inequalities from the start and redesign mentoring relationships to include groups and to take advantage of Gurin's recommendations for structuring "safe" dialogues. These include creating relatively balanced groups in terms of numbers of senior and junior faculty (and if possible a critical mass of junior faculty who are from backgrounds underrepresented in the discipline or department). She also recommends mentoring by group, rather than one-on-one, to allow for intra-group consensus and affirmation, and intergroup differences (in perspectives, lifestyles, opinions), to emerge. Each group should have sufficient variability to break down the notion that the group is monolithic in its opinions while building an appreciation of intergroup similarity and common cause.

Although some departments may consider themselves too small to constitute such groups, there is nothing to stop them from coming together to form mentoring groups. In fact, in light of the interdisciplinary trends in many fields from the arts to the sciences, this may be especially useful. The same instrumental passing on of advice could be achieved, but this time it might be received more as a revelation about how things have worked than as a demand for conformity in the future. The consensus support coming to the junior faculty from sharing their experiences with other colleagues in similar positions could be very affirming. As a consequence, defensive barriers on both sides could erode, especially if faculty evaluation were separated from the intergroup mentoring dialogues, building the basis for common cause.

The democratic culture built on intra-group affirmation and healthy intergroup exchange opens everyone up to healthy self-examination and constructive exploration. The university benefits from new voices and perspectives, and the newer entrants into the professoriate get more transparent access to the pathways to power and longevity in the institution.

Connecting Universities and Communities in Democratic Culture

In practical ways, universities are good places to build democratic cultures because the freedoms of the academy allow for the airing of difference and conflicts in relatively civil settings. This does not mean that universities are immune to societal disparities and divisions, but it does suggest that they allow for more than the usual experimentation with "difficult dialogues." However, universities are also at risk of stultifying insularity and the self-righteousness it breeds if they do not open themselves up to—and look out at—the world beyond the campus.[27] Of course, building democratic culture between campus and community is no easier than it is on campus, although just as important. In my view, for these efforts at engagement to be authentic, the issues and partnerships and dialogues should arise "organically" from the specific intellectual, cultural, and historical landscape common to both campus and community. In other words, it works best when it "belongs" at least somewhat to the identity/culture of both "groups."

I made this argument when analyzing the success of the University of Michigan in waging its defense of affirmative action.[28] Defending diversity grew somewhat organically out of the historical and contemporary landscape of this great public institution. The university had a long and storied (if not always peaceful) history of activism around issues of race, intergroup dynamics, and social research. Its location near Detroit, a Northern icon in the struggle for civil rights and a contemporary reminder of the ills of White flight, provided authenticity to the urgency of its fight. As did the competitive concerns (for educating a diverse and culturally competent workforce) on the part of the major industrial and labor organizations of this Big Three-dominated state. No wonder there was a surprisingly strong (though nowhere near uniform) appetite for taking on this nationally contentious issue in a protracted and expensive defense of affirmative action in college admissions.[29]

The same can be said for some of the best work that Syracuse University is pursuing today. Syracuse, New York, and its surrounding region have an historic legacy as an arena in the struggle for the rights of women, slaves, and Native Americans. As the author Charles C. Mann noted several years ago in *The New York Times*,[30] our region is the ancestral home of the Onondaga Nation and the capital of the historic Confederacy, which was "probably the greatest indigenous polity north of the Rio Grande in the two centuries before Columbus and definitely the greatest in the two centuries after."[31] "Haudenosaunee" translated into English means "People of the Longhouse" and includes the Mohawk, Oneida, Onondaga, Cayuga, Seneca, and Tuscarora Nations. Their alliance was governed by a constitution, the Great Law of Peace, which established the league's Great Council, with 50 male religious-political leaders, each of whom represented the female-led clans of its member nations.[32] Under the Great Law, governance by the Great Council focuses on a balance of power and consensus that puts first the well-being of the people. In all matters, the Haudenosaunee look ahead to generations yet unborn to ensure that the same good life and natural world that exist today survive for future generations. According to ethnographer Lewis Henry Morgan, the Great Law sought to avoid the concentration of power in the Council as much as it gave it authority, and it also sought to prevent power from falling into the hands of any single individual.[33]

As Mann wrote in the *Times*, the Indigenous government was predicated on the consent of the governed and "compared to the despotisms that were the norm in Europe and Asia, the societies encountered by British colonists were a libertarian dream." Many European settlers found in them a "deeply attractive vision of human possibility," and such authors as Locke, Hume, Rousseau, and Thomas Paine took from them many of their examples of liberty. Even the suffragists Elizabeth Cady Stanton and Matilda Joslyn Gage, who lived in the Finger Lakes region, were inspired by the Great Law's legal protections to women.[34]

Syracuse University is situated in the historic capital of this great and ancient confederacy and has more than 250 Indigenous alumni. One of them, Robert Odawi Porter, a member of the Seneca Nation (Heron Clan), law professor, and consultant to several Indigenous Nations and organizations, founded and directs our Center for Indigenous Law, Governance and Citizenship in the College of Law. This research-based law and policy institute focuses on Indigenous Nations, their development, and their interaction with the U.S. and Canadian governments. Chief Sidney Hill, the Tadadaho, or spiritual leader, of the Haudenosaunee, was invited to participate in the

Chancellor's inauguration in 2004, and a number of Clan Mothers also accepted the university's invitation to join the academic procession. Chief Hill addressed the gathering in the Onondaga language, the first time many of us non-Natives had ever heard it. Translating for the audience was Faithkeeper Oren R. Lyons, a Syracuse University (SU) alumnus and distinguished professor in the Department of American Studies at the University at Buffalo, State University of New York. In November 2007, Syracuse University formally dedicated Lyons Hall, the former International Living Center, in his honor.

Two years ago, the university and the Onondaga appointed representatives to collaborate on a new, mutually beneficial and sustainable relationship, including a program we are calling the Haudenosaunee Promise that provides full scholarships and expenses for Haudenosaunee first-year and transfer students admitted to SU and residing in one of the Six Nations' territories. The program has set all-time records for the size of the applicant pool and the enrollment of well-qualified Native students. At the same time, we have enhanced our course offerings in the Native American Studies Program and established major new advocacy and research centers. A year-long lecture series, organized by Neighbors of the Onondaga Nation and sponsored by many departments at Syracuse and the State University of New York College of Environmental Science and Forestry, drew huge audiences from both campus and community to hear Indigenous leaders discuss "Onondaga Land Rights and Our Common Future" and is continuing in 2010.

Indeed, the notion of a common future must be the foundation on which we build our efforts. Syracuse University is hoping for a wide range of collaborations with the Onondaga and other members of the Haudenosaunee, partnerships that are organic, not only in proximity and the history of our region, but also in our scholarly and educational interests, including those of our Departments of Religion, Women's Studies, and African and African-American Studies, and such shared environmental justice interests as cleaning up Onondaga Lake and improving the urban ecosystem of our region.

Authenticity—that is, committing to things that are organic to one's group or institution or community—is, in my view, at the core of success in building democratic culture, on campus and with connected communities beyond. Such engagements have immediate credibility and potential for broader societal impact, lending support to the role of universities, private as well as public, as public goods.

Moving Closer; Crossing Boundaries

Many fault lines divide and paralyze peoples in today's world. They keep us apart socially and intellectually, and they stymie access to opportunity and mobility for those who fall on the wrong side of one or another line. In an increasingly diverse world, we must recognize that groups have a vital role to play, providing not only the richness of a multicultural world but a healthy sense of place and affirmation for individuals. If we are to reap the full benefits of this diversity, however, we must do the work of building democratic culture, and it involves crossing fault lines to move together, not apart. Therefore, we must undertake not only such programmatic efforts as intergroup dialogue curriculum, faculty mentoring, and campus-community collaborations, but also be vigilant about the physical landscape of our activities—the places and spaces in which we do our work and the freedom we feel to cross boundaries and explore. We have to ask in very concrete terms whether we are *moving together*. A large part of why the intergroup work is so hard is that most of us have precious little experience living and working and going to school together. That is why the Civil Rights Project (originally called the Harvard Civil Rights Project) has cemented so much of its analysis in explicating patterns of residential segregation, for example.[35] And, that is why Syracuse University, a campus quite literally poised on a hill and divided by Interstate 81 from the city's quite remarkable cultural offerings and its resilient inner-city neighborhoods, has committed itself to an agenda called "Scholarship in Action."

Scholarship in Action is grounded in the belief that we will open up rich new opportunities for learning, innovation, and discovery if we can create settings where we are deeply engaged with each other and with "communities of experts" on and off our campus. Cross-disciplinary collaboration can be invaluable in addressing critical societal issues—whether we are talking about environmental sustainability, failing schools, or shrinking cities. At the outset, we purchased and renovated an old furniture warehouse downtown to house our School of Architecture, design programs, a new graduate program in arts journalism, public galleries, and meeting rooms for area artists. The Warehouse anchors an urban pathway we are calling the Connective Corridor, a public-private partnership in community development and design. It will carve out an arts district from our campus on the hill into downtown and on to our Near Westside, a poor and largely minority neighborhood. This zone for the arts, design, and technology includes historic churches, community art spaces, performance spaces, and museums and has created wonderful collaborations at every level. It is being designed to respect community, enhance the urban landscape, and increase accessibility for people with disabilities.

We have also set up centers for entrepreneurship, literacy, and the arts in city neighborhoods, especially to empower and develop the South Side of Syracuse, which has one of the highest child poverty rates in the state of New York. These are joint efforts with community groups, foundations, nonprofit organizations, and city and state governments. Our Whitman School of Management has spearheaded the South Side Innovation Center for women- and minority-owned business start-ups, our I-School is spreading centers of wireless technology in the neighborhood, our College of Law is partnering to establish a food co-op here, and our SI Newhouse School of Public Communications is working with residents to establish a local newspaper. We are engaged with other institutions of higher education in Syracuse in a Partnership for Better Education with the Syracuse City School District that includes all 21,000 children in every school in the city and has major new initiatives in literacy, science, and the arts, as well as new programs to encourage teaching in high-needs fields. We are also leading a multi-institution university-industry consortium on Environmental Systems and Energy, with a state-funded headquarters down from University Hill, adjacent to where the old Erie Canal served as the engine for innovation in an earlier era.

Universities around the country are engaging with their connected communities as we are, and these efforts, especially when they involve the best academic programs and derive authentically from the expertise and concerns of the campus and the community, are only likely to increase. As this happens, and if it is sustained and complemented by recruitment on campuses of diverse faculties and student bodies, then people and ideas may well traverse those otherwise impermeable fault lines of intergroup relations, on campus and beyond. If so, then we can say that democratic culture is taking hold.

Notes

1. Now the Civil Rights Project at UCLA.
2. Erica Frankenberg, Chungmei Lee, and Gary Orfield, "A Multiracial Society with Segregated Schools: Are We Losing the Dream?", The Civil Rights Project, UCLA (January 16, 2003) 28, http://www.civilrightsproject.ucla.edu/research/reseg03/resegregation03.php (accessed January 23, 2010).

3. Isaiah Berlin, "The Crooked Timber of Humanity," in *Chapters in the History of Ideas*, ed. Henry Hardy (New York: Vintage Books, 1992), pp. 18–19.
4. Patricia Gurin, Ratnesh Nagda, and Gretchen Lopez, "The Benefits of Diversity in Education for Democratic Citizenship," *The Journal of Social Issues* 60.1 (2004): 17–34.
5. Patricia Gurin, E. Dey, S. Hurtado, and G. Gurin, "Diversity and Higher Education: Theory and Impact on Educational Outcomes," *Harvard Educational Review* 72.3 (2002): 330–366. Also, Nancy Cantor, "Higher Education Policy-Making in the Melting Pot of Stakeholder Voices: The Michigan Affirmative Action Cases," *Policy Forum, Institute of Government and Public Affairs* 17.2 (2004), http://igpa.uillinois.edu/system/files/PF17-2.pdf (accessed January 23, 2010).
6. P. Brickman and D. Coates, "Commitment and Mental Health," in *Commitment, Conflict, and Caring*, ed. P. Brickman (Englewood Cliffs, NJ: Prentice-Hall, 1987), pp. 222–309. R. F. Baumeister and M. R. Leary, "The Need to Belong: Desire for Interpersonal Attachments as a Fundamental Human Motivation," *Psychological Bulletin* 117.3 (1995): 497–529.
7. N. Cantor and C. Sanderson, "Life Task Participation and Well-Being: The Importance of Taking Part in Daily Life," in *Well-Being: The Foundations of Hedonic Psychology*, ed. D. Kahneman, E. Diener, and N. Schwarz (New York: The Russell Sage Foundation, 1999), pp. 230–243. Also Amartya Sen, *On Ethics & Economics* (Cambridge: Blackwell, 1994), pp. 43–45, recognizing the distinction between the "agency aspect" and the "well-being aspect" of a person.
8. N. Cantor, M. Kemmelmeier, J. Basten, and D. Prentice, "Life Task Pursuit in Social Groups: Balancing Self Exploration and Social Integration," in *Self and Identity*, ed. C. Morf and W. Mischel, Special Issue, 1 (2002): 177–184.
9. Patricia Gurin, *Educational Benefits of Intergroup Dialogue*, unpublished grant proposal, University of Michigan, Ann Arbor, MI, August 2004.
10. Many of these insights come from conversations with Professor Gretchen Lopez, who directs Syracuse University's participation in the Multiversity Intergroup Dialogue Project.
11. Ford Foundation call for proposals: "Difficult Dialogues Initiative: Promoting Pluralism and Academic Freedom on Campus," July 2005.
12. Jodi Wilgoren, "In One Prison, Murder, Betrayal, and High Prose," *The New York Times*, Friday, April 29, 2005, Section A, 16.
13. Khaled Hosseini, *The Kite Runner* (New York: Riverhead Books, 2004), p. 9.
14. Ibid., 9.
15. Ibid., 9–10.
16. Nancy Cantor, op-ed in *Chicago Sun Times*, June 22, 2003, responding to Chetly Zarko, "The Evidence of Things Not Seen," *The Wall Street Journal*, May 16, 2003; Nat Hentoff, "Affirmative Action Discords," a

column that appeared in the *Jewish World Review*, May 28, 2003; and Arthur Levine and Jeanette Cureton, *When Hope and Fear Collide: A Portrait of Today's College Students* (San Francisco, CA: Jossey-Bass Publishers, 1998).
17. Derek V. Price and Jill K. Wohlford, "Equity in Educational Attainment; Racial, Ethnic, and Gender Inequality in the 50 States," in *Higher Education and the Color Line*, ed. Gary Orfield, Patricia Marin, and Catherine L. Horn (Cambridge: Harvard Education Press, 2005), p. 65.
18. Expert Report of Claude M. Steele, *Gratz* and *Grutter*, in the U.S. District Court for the Eastern District of Michigan.
19. Patricia Gurin, *Educational Benefits of Intergroup Dialogue*, unpublished grant proposal, University of Michigan, Ann Arbor, MI, August 2004.
20. M. C. Thompson, T. G. Brett, and C. Behling, "Educating for Social Justice: The Program on Intergroup Relations, Conflict, and Community at the University of Michigan," in *Intergroup Dialogue: Deliberative Democracy in School, College, Community, and Workplace*, ed. David Louis Schoem and Sylvia Hurtado (Ann Arbor, MI: University of Michigan Press, 2001), pp. 99–114.
21. David Horowitz, "In Defense of Intellectual Diversity," *The Chronicle Review, The Chronicle of Higher Education*, February 13, 2004.
22. Personal communication with Gretchen E. Lopez, director of the Syracuse University-Multiversity Intergroup Dialogue Project, November 2007.
23. For more information, see http://www.irccny.org/programs/cwd/cwd_about.php.
24. For more information on Syracuse University's intergroup dialogue programs see http://intergroupdialogue.syr.edu/AcadCourses.htm.
25. American Council on Education, Office of Women in Higher Education Report, "An Agenda for Excellence: Creating Flexibility in Tenure-Track Faculty Careers," February 2005. See also Nancy Cantor and Steven D. Lavine, "Taking Public Scholarship Seriously," *The Chronicle of Higher Education*, June 9, 2006, B20.
26. Julie Ellison, Director's Column, *Imagining America Newsletter* (Summer 2004). Also, Nancy Cantor, "Valuing Public Scholarship," *The Presidency: The American Council on Education Magazine for Higher Education Leaders* (Spring 2005): 35–37.
27. Nancy Cantor and S. Schomberg, "Poised between Two Worlds: The University as Monastery and Marketplace," *Educause Review* 38.2 (March/April 2003): 12–21.
28. Nancy Cantor, Introduction to: *Defending Diversity Affirmative Action at the University of Michigan*, ed. P. Gurin, J. Lehman, and E. Lewis (Ann Arbor, MI: University of Michigan Press, 2004), pp. 1–16.
29. The overthrow of affirmative action at Michigan through a ballot referendum in November 2006 may illustrate how the economic woes of a

state can encourage a zero-sum pitched battle for access and overwhelm the positive context for civil rights coalitions.
30. Charles C. Mann, "The Founding Sachems," op-ed in *The New York Times*, July 4, 2005, A13.
31. The term "Iroquois," by which most of us know this confederacy, was concocted by the French as an insult meaning "black snake."
32. Mann, "The Founding Sachems."
33. Ibid.
34. See Matilda Joslyn Gage, *Woman, Church and State*, ed. Sally Roesch Wagner (Aberdeen, SD: Sky Carrier Press, 1998 [1893]), pp. 5–6, and Sally Roesch Wagner, *Sisters in Spirit: Haudenosaunee (Iroquois) Influences on Early American Feminists* (Summertown, TN: Native Voices, 2001).
35. *Higher Education and the Color Line*, ed. Gary Orfield, Patricia Marin, and Catherine L. Horn (Cambridge: Harvard Education Press, 2005), p. 65.

2

From Diverse Campuses to Integrated Campuses: How Can We Tell if We Are "Walking the Walk"?[1]

*Jeffrey S. Lehman**

The Supreme Court's decision in *Grutter v. Bollinger*[2] affirmed the authority of universities to employ admissions policies that provide a broadly diverse community of qualified applicants with access to their campuses. In a thoughtful majority opinion by Justice Sandra Day O'Connor, the Court held that it does not automatically violate the Equal Protection Clause if public universities deliberately work to create communities that include pedagogically meaningful numbers of students from a broad array of racial, ethnic, religious, socioeconomic, and ideological backgrounds.

I was a named defendant in the litigation, since I was the dean of the University of Michigan Law School at the time. And people who are friendly to our position sometimes pose questions such as the following:

"Why was that so difficult?"
"Why was it only a 5–4 victory, rather than a 9–0 decision?"

---

* Jeffrey S. Lehman is Professor of Law and former President of Cornell University, currently on leave to be Chancellor and Founding Dean of the Peking University School of Transnational Law. He previously served as Dean of the University of Michigan Law School, where he was one of the architects of the school's successful defense of its admissions program before the U.S. Supreme Court. His scholarship deals with the poverty and inequality in the American welfare state.

"Why were people so angry with you for promoting such an obviously worthwhile goal?"

I have been thinking about questions such as these for many years, and I believe the answer is clear. Whenever we employ affirmative action, we are using a very dangerous tool. Think, if you will, of a very sharp knife. Or a caustic chemical. Or perhaps a technology like recombinant genetics. Think of any tool that could, in the wrong hands, cause enormous harm to innocent people.

The tool here is the category of race. Through affirmative action, we are using race to do more than just describe the world we see around us. We are giving race performative significance. We are looking at individual applicants, classifying them according to race, and then using the result of that classification process as a factor in how we allocate valuable opportunities.

America's history tells us that this is a very sharp knife indeed. For almost four hundred years, we have seen racial categories used to construct systems for the subordination of individuals who were assigned to disfavored groups. And it is important to remember that across those four centuries the people who were using the knife invariably believed they were doing so in order to promote an important social end.

The constitutional jurisprudence of the Equal Protection Clause and the legislation of the civil rights era have given us an approach to the use of race as a category by large and powerful institutions. Our society does not ban the knife. It does not outlaw the technology. It does not say that such institutions must always act in a rigidly colorblind fashion.

Instead, our legal system has chosen to rely on the concept of a rebuttable presumption. It declares this kind of classification to be "suspect." Recognizing that racial classifications can do enormous harm in the wrong hands, it holds their use up to "strict scrutiny." Our system declares their use to be presumptively illegitimate, permissible only if the institution that wishes to use them can show that its actions are "narrowly tailored" to promote a "compelling interest."

In the *Grutter* lawsuit we were able to clear that hurdle. Justice O'Connor's opinion affirmed that our society has a compelling interest in creating university campuses that are meaningfully integrated. That interest is compelling because an integrated campus promotes certain learning outcomes. And it is compelling because the legitimacy of our democracy depends upon the existence of open, visible paths to leadership for people of all races who are talented and qualified.

Justice O'Connor's opinion also affirmed that the law school's admissions policy was narrowly tailored to promote that interest. Because we were being careful with the knife, we would be allowed to keep using it.

So why was it so difficult? Why did the other eight justices not agree with Justice O'Connor? And why, in 2006, did the voters of Michigan adopt Proposition 2, a ballot initiative saying that they would prefer university admissions to be rigidly color-blind?

I believe a significant part of the answer has to do with trust. Not everyone trusts universities with sharp knives. Not everyone believes that universities are sincere. One need look no further than Justice Scalia's vituperative dissent in *Grutter*, in which he refers to "universities that talk the talk of multiculturalism and racial diversity in the courts but walk the walk of tribalism and racial segregation on their campuses." The charge is hypocrisy—or, more gently, an incapacity to follow through on our commitments.

If we want the larger political community to trust universities with the dangerous category of race, it is crucially important that they see follow-through from the universities that are still permitted to employ affirmative action. Universities must show that they are committed to reaping the benefits that justified the use of affirmative action in the first place.

How can a university "walk the walk"? Once the admissions process has produced a diverse community, how can it ensure that the community functions well? How can it ensure that campus diversity leads to a flourishing, integrated learning environment that is characterized by curiosity, civility, and a shared commitment to understand and appreciate the complex truths that define our world?

The first step is to take seriously the project of self-evaluation.

Today's debates and discussions about whether universities follow through on their declared commitments to diversity are predictable and superficial. Critics point to lunch tables in the cafeteria and say, "You're balkanized, you're segregated." Defenders point to classrooms and say, "We're integrated."

We can do better than that.

I would begin by being very precise about what pedagogic benefits we hope to obtain through integrated campus communities. To be fair, there is no consensus on what those benefits should be. So an evaluation project must begin with a willingness to make choices, to declare precisely what goals a given institution has in its collective mind.

For purposes of discussion, I focus my attention on two potential benefits of integrated campuses that are, it seems to me, worth fighting for.

First is the intellectual benefit of *perspective enlargement*. Bowen, Kurzweil, and Tobin say, "Students need to learn how to put themselves in other people's shoes."[3] Others might say that students need to see the world through other people's eyes. In speaking to students I often invoke Keats's notion of "negative capability," the ability to entertain two opposing ideas "without irritable reaching after fact and reason." These are all different approaches to transforming ourselves, intellectually, imaginatively, and emotionally. Sometimes we transform ourselves by crossing a boundary, from one side to another. Sometimes we transform ourselves by levitating, so that we can see both sides of the boundary at the same time.

The point here is that, on our way to becoming better at *imagining* multiple perspectives, it is enormously helpful to build a base of experience by *listening* to others describe how they actually see a situation. It is helpful to listen through the voice of literature. It is also helpful to listen to a live person, someone you know and trust, who is sharing the same experience with you and who is prepared to engage in sustained conversation that goes beyond a quick exchange.

The second benefit I would emphasize is, in some ways, a special case of the first. It is the particular experience of *role reversal* associated with one's identity as a member of a minority group or as a member of the majority. Because of residential segregation, very few of our white students have had a chance before college to experience what it feels like to be, racially, in the minority. Our minority students are, of course, much more likely to have experienced both contexts in which they were members of the predominant group and contexts in which they were not. But most of them had secondary school experiences that were mostly one way or the other—either they grew up as part of a relatively small minority in an overwhelmingly white environment or they grew up in an environment that was heavily "majority minority."

How can we tell if, overall, students on our campuses are getting the benefits of perspective enlargement and role reversal?

I believe that there is no better place to start than in the much-mooted campus cafeteria. Does the existence within a cafeteria of significant numbers of racially homogeneous lunch tables mean that the project of integration is a failure? Does the existence of dormitories or fraternities that are majority-minority mean the same thing?

Not necessarily.

These phenomena certainly prove that universities are not raceless communities. They are not color-blind. But that should be no surprise; everyone notices race. Those who observe the cafeterias notice race, and the students who are being observed notice race as well.

Moreover, race remains more than just something to be noticed, like hair color. Race continues to have substantial social meaning. It is sufficiently important that being in an integrated environment carries with it the kind of opportunities for perspective enlargement and role reversal that I described earlier.

But the very circumstances that create those opportunities also mean that an individual who wants to capture them must accept a certain degree of effort, of stress, of social risk. It is entirely understandable that people might want, at times, to opt for lower-stress environments. Our students do not need to be always distributed randomly in order to benefit from an integrated campus.

What, then, might be a fair measure of success, short of round-the-clock random distribution? Here, I would return to a concept that figured significantly in the University of Michigan Law School's admissions policy, the concept of "meaningful integration." I would like to suggest that on a meaningfully integrated campus, *most* students are having a meaningful experience of perspective enlargement and role reversal.

As I envision it, a meaningfully integrated campus is characterized by behavior at the level of the individual, and a culture at the level of the university. The individual experiences a daily ebb and flow between people like oneself and people who are different. The individual makes a commitment to move back and forth every day between the spaces that are safe and the spaces that nurture growth and challenge. The university in turn makes a commitment to value and support that kind of movement back and forth by its members. The university expressly endorses the importance of being meaningfully integrated—in action and not just in demographics.

If that is what it means to be meaningfully integrated, at the level of behavior and at the level of culture, how will we know it when we see it? The question is much more difficult than it might sound.

Imagine a community of 100 students, all of whom are either orange or green. Suppose 80 are orange and 20 are green. And suppose that no matter when you looked at the community you saw the same thing: 75 orange students huddled together in Place A, 15 green students huddled together in Place C, and one "actively integrated pod" (Place B) with 5 orange and 5 green students. Such a community might be diagrammed as follows:

	A	B	C
Orange	75	5	
Green		5	15

I would submit that you do not yet have enough information to say whether the overall community is meaningfully integrated.

To be sure, if it were the same 5 orange students and the same 5 green students in Place B, every hour of every day, then one could readily conclude that the community is failing to be meaningfully integrated, for 90 percent of the students would not be experiencing the benefits of integration.

However, one might reach a very different conclusion if one knew that, over the course of the day, there were a constant rotation of students in and out of Place B. For example, suppose that, within any given 16-hour day, each of the 20 green students is spending 4 hours in the integrated sector. And suppose that 60 of the 80 orange students are each spending 1 hour and 20 minutes in the integrated sector. Such a community might be diagrammed as follows:

	A	B	C
Orange	20 (55	⇔ 5)	
Green		(5 ⇔	15)

It would not be difficult to conclude that each of those 80 students is experiencing a meaningful ebb and flow in their lives. Only 20 percent of the students (all orange) are getting no integrated experience at all. One might well decide that this community is meaningfully integrated in the sense in which we have been using the term.

This stylized example suggests that the evaluation project should be structured with the following points in mind.

First, it is important to know what phenomena are significant. One should not feel any particular sense of alarm if, when one looks into a lunchroom, one sees a table at which all of the students are black. What one needs to look for are integration pods—venues where different race groups are interacting. Even a relatively small number of such venues *might* be enough to sustain a meaningfully integrated community.

More precisely, one needs to be looking at *flows*, not snapshots. An evaluation should be structured to monitor the movements within an individual's daily life. It should be longitudinal rather than cross-sectional.

It would probably also be useful to recognize two different kinds of integration pods, which offer two different kinds of benefits. Pods that include meaningful numbers of students from different races are where one might expect to find perspective enlargement. If there are meaningful numbers of students from each race, we are more likely to find multiple perspectives expressed *within* each race, so that students from other races acquire more nuanced understandings of the extent to which race does and does not shape perspective. These kinds of integration pods might be contrasted with "role reversal pods," where a handful of white students mix with a larger number of minority students. Such pods do a special kind of educational work, one that holds distinctive value for a university.

Next, it is important to show care when evaluating data about situations where individuals spend different proportions of their time inside and outside the integrated sector. A helpful analogy may be found in the social science literature concerning the dynamics of poverty spells. The seminal article on the topic was written two decades ago by Mary Jo Bane and David Ellwood.[4]

Bane and Ellwood considered the question, "Is the poverty problem primarily one of people cycling quickly in and out, or is it primarily one of people staying stuck in poverty for a long time?" One might paraphrase their argument by reference to the following stylized example. Suppose that, over the course of a nine-year span, each of nine different people $(a \ldots i)$ is poor for exactly one year, with poverty striking a different person every year. Suppose also that a tenth person (j) is in poverty throughout the entire nine-year span. A diagram might look like this:

```
1    a    )
2    b    )
3    c    )
4    d    )
5    e    ) j
6    f    )
7    g    )
8    h    )
9    i    )
```

Within the overall group of ten people, only one of ten (person *j*) is poor for a long spell. Thus, one might say that the incidence of long-term poverty is 10 percent. On the other hand, in any given year, or indeed at any given point in time, two people would be poor: person *j* and one of the other nine. Looking from this perspective, one might say that the incidence of long-term poverty is 50 percent (at any given moment, 50 percent of those who are poor are in the midst of a long-term spell). Is long-term poverty half the problem or a tenth of the problem? It depends on whether you are interested in "poor now" or "ever poor."

Consider once more the context of racial integration. Suppose one were to take a series of snapshots of a role-reversal integration pod. Suppose that in every snapshot one found eight black students and two white students. And, upon closer inspection, suppose one found that one of the two white students was always the same.

It might be tempting to focus on the student who was always there and say, "This integration pod isn't having much of an impact; the same white student is here all the time." But if the point is to evaluate the prevalence of a role-reversal experience within the community over time, the lesson of the poverty spells example is that one must not look at individual snapshots. One must look at the entire population over time. The question is not "there now," but rather "ever there."

There is a common point to both the cafeteria example and the role-reversal example. The point is that, to evaluate how one is doing, one must resist the psychological temptation to fixate on indicators of failure. One must not fixate on homogeneous lunchroom tables or on the white student who seems to be in the role-reversal pod every day. Rather, one must try to see the entire picture, over time.

Thus, to understand how well a university is pursuing integration, there is no substitute for serious empirical investigation. I do not know of any university that is taking up such investigation. I hope one does soon. And I hope that it reports its results to the larger community.

And after conducting the investigation, what then? It depends, of course, on what the investigators reveal. But I feel comfortable speculating that, on most American campuses today, a rigorous investigation will reveal a story of mixed success. It will likely find a small but growing number of students who are fully reaping the benefits of living in an integrated environment. It will likely find a small and shrinking number of students who spend four years in a diverse campus and derive almost no benefit from having done so. Finally, it will likely find a large, heterogeneous middle group that seems to be getting some benefit, but perhaps not as much as we might wish.

What should an academic leader do if presented with findings such as these?

One response might be to do nothing, especially if it appears that things are getting steadily better. And there are certainly reasons to think that things *are* getting better. My own impressionistic sense of America's campuses is that students today are living significantly more integrated lives than they did 20 years ago. The emergence of hip-hop as a central feature of youth culture has changed the terms of engagement of students across racial boundaries. So has the rapid emergence of strong Latino and Asian presences within our community, catalyzing a shift in our self-understanding from a biracial model to a multiracial model.

So, if things are getting better, inaction might be tempting. After all, the category of race is a very sharp knife, and the larger society counsels faculty and administrators to wield it with caution. In addition, we know that students have somewhat ambivalent feelings about whether they want to follow faculty leadership or rebel against it.

And yet, as tempting as inaction might be, I think that faculty and administrative leaders can safely do at least a little more. In my experience, as long as they are not too heavy-handed, faculty members can have an impact on the culture of a campus. They can gently but effectively nudge their students in the direction of a daily ebb and flow. And they can nurture the integration pods they see on campus.

As president of Cornell, I was struck by the extraordinary alliance between Muslim students and Jewish students on campus. Year in and year out, I saw remarkable efforts at joint programming, so that students from each group might come to appreciate the perspectives of the other. One year the students collaborated to design and make a mosaic. In other years the end of Ramadan featured an Iftar banquet that drew hundreds and hundreds of students into a comparative discussion of the role of fasting in different religions. In still another year Iranian Muslim students and Sephardic Jewish students staged a joint evening of culture, food, and comedy.

These were important integration pods. They emerged from student ideas. But when I spoke with the student leaders who brought them about, they impressed on me how critical they felt it was that faculty members endorsed them. Not so much through public pronouncements, but by showing up themselves and participating. I do not think these students were just saying it to make me feel that faculty are important. I think they were sincere.

In summary, it is important for many reasons that students at universities that practice affirmative action in admissions actually reap the benefits of those practices. Campuses should be not merely diverse but also integrated. To prepare themselves for life in an integrated world, the students should take advantage of the opportunities for perspective enlargement and role reversal. It is unrealistic to expect them to do so all the time; it is a worthy aspiration that they will experience a daily ebb and flow between the comfortable and the challenging.

Universities should evaluate how successful they are in this regard. They should do the work of evaluating the prevalence of integration pods on campus, and the extent to which they touch the lives of a broad cross-section of our students. And they should nurture them as best they can.

If a university does all these things, then it will have used the sharp knife of integration to carve a work of great importance. It will be walking the walk of integration in a way that even Justice Scalia should admire.

Notes

1. This essay was prepared as a paper for presentation at a symposium entitled, "Diversity and Excellence in American Higher Education: The Road Ahead," at Cornell University, July 30, 2005.
2. *Grutter v. Bollinger.* 539 U.S. 306 (2003).
3. William G. Bowen, Martin A. Kurzweil, and Eugene M. Tobin, *Equity and Excellence in American Higher Education* (Charlottesville, VA: University of Virginia Press, 2005), 4.
4. Mary Jo Bane and David T. Ellwood, "Slipping Into and Out of Poverty: The Dynamics of Spells," *Journal of Human Resources* 21.1 (1986): 1–23.

3

IS DIVERSITY WITHOUT SOCIAL JUSTICE ENOUGH?

Michael Hames-García*

INTRODUCTION

I recently returned to the small liberal arts college where I received my B.A., not to teach and not to attend classes or an alumni event. Instead, I was returning to give advice to faculty in my former major department concerning its historical challenges recruiting and retaining faculty of color. One of my former professors thought that, as an alumnus of color who directs an Ethnic Studies program (now an Ethnic Studies department) at a large, nearby state university, I might be able to offer some useful insights about how to further the college's and the department's diversity goals. One consequence of returning to my alma mater was a series of reflections about what "worked" for me, what about this college from my perspective as a former student helped me to stay in college. What, specifically, made remaining at *this* college seem doable—particularly given that there were no U.S.-born faculty of color in arts and sciences at the time that I was there and given that there were pitiably few students of color (as far as I know, there were two Latinos in my entering class on a campus of around one thousand students)?

This visit coincided with a request from the editors of this volume to write on the topic of equity and diversity in higher education.

---

*Michael Hames-García has been Program Director and Department Head of Ethnic Studies at the University of Oregon since 2006. He also directs the Center for Race, Ethnicity, and Sexuality Studies (CRESS) at the University of Oregon. He is a founding member of the Future of Minority Studies Research Project (FMS) and the author and editor of several books.

In short, I was asked, based on my experience—as an undergraduate at a small, private liberal arts college, as a graduate student at a large, Ivy League research institution, as a junior faculty member in a traditional humanities discipline at a large, highly diverse public research institution, and as director of an Ethnic Studies program at a significantly less diverse public research institution—what I think works in improving equity and diversity for underrepresented groups in higher educational institutions. While I believe that pipeline issues and the dire state of primary and secondary education for people of color in the United States are probably the biggest obstacles to diversifying our colleges and universities, I have arrived at three related recommendations specific to institutions of higher education. I believe that some of these recommendations can also spill over to addressing the issues around the larger pipeline of primary and secondary education, as well.

The first recommendation is to better integrate the university's mission of scholarly excellence into "multicultural affairs" agendas, breaking down the mighty barrier between student affairs and academic affairs and thereby infusing multicultural affairs initiatives with greater intellectual weight and legitimacy in the academy. Second, I believe that it is vital for equity and diversity initiatives (including curricular initiatives) to view any commitment to diversity as part of a larger commitment to social justice, rather than as an end in itself. Indeed, it might be advisable to discard the language of "multiculturalism" and "diversity" in favor of more expansive commitments to social justice. Finally, I would like to recommend ongoing encouragement (rather than a mere tolerance) of the sometimes-inconvenient student organizing that disrupts business as usual on our campuses and frequently forces us to reexamine our moral values and intellectual commitments.

Can Diversity Be a Substantial Part of the Academy's Research Mission?

The common wisdom holds that a surefire way to recruit and retain more students and faculty of color is by recruiting and retaining more students and faculty of color. This strategy usually goes by the name of "achieving critical mass." I agree with it, but with the increasing restrictions on affirmative action practices and the difficulties in increasing the number of people of color in the faculty pipeline, we have to have broader strategies as well. Until an institution can achieve "critical mass," how can it retain students and faculty who might feel isolated on an overwhelmingly white campus? Student Affairs offices are always involved in decisions about where and how students of color spend their time and receive their support. But what relationship should exist between student affairs and faculty, particularly faculty of color? As chair of an Ethnic Studies department, I am acutely aware of the gulf that exists between research and academic affairs, on the one hand, and institutional equity and diversity and student affairs, on the other. To be blunt, most research faculty, including faculty of color, have a primary identification with the research and academic mission of the university rather than with the kinds of work faculty typically associate with student affairs and offices of equity and diversity. Institutional climate surveys, workshops about stereotypes and overcoming prejudice, and similar activities usually directed toward undergraduates commonly do not seem to be directly engaged with the kind of research that faculty in Ethnic Studies and women's and gender studies, broadly speaking, see themselves doing. As a result, administrators and staff in diversity offices and student affairs are (often unfairly) viewed as intellectually "light" when it comes to precisely those subjects that are most important in recruiting and retaining faculty and students of color.

Those faculty members who enter administrative positions in student affairs or equity and diversity risk a loss of prestige. Ethnic Studies faculty sometimes see such administrators as having given up on real research in order to engage in feel-good activities aimed at "celebrating our differences," rather than addressing legacies of racism and ongoing conditions of structural inequality. At other times, they view them as gatekeepers whose job it is to run interference for an administration resistant to real change. In her May 2008 address to the National Conference on Race and Ethnicity in Higher Education, Evelyn Hu-DeHart accused those working on diversity in higher education with "covering up" for colleges and universities that are not really interested in addressing racism. According to *The Chronicle of Higher Education*,

> [Hu-DeHart] asked her audience to comb through the program for the five-day meeting and note the job descriptions of those who would be speaking, and think about those who seemed absent from this event. The group found plenty of listings for chief diversity officers, administrators and staff members from campus offices in charge of

student support, outside diversity consultants, and faculty members in the fields of education, psychology, and ethnic studies. But they found little evidence of the presence of college trustees, presidents, provosts, academic deans, or professors in more traditional academic fields, especially mathematics and science.

Many of those missing, she said, are "the heart of the academic side" of colleges, people who have power over research, curriculum, and the hiring and evaluation of faculty members.[1]

Regardless of the fairness of such criticisms or the good intentions of diversity administrators, the gap between them and research faculty is real, and it continues to be, in my mind, one of the most important barriers to making recruitment and retention schemes truly effective. So long as research faculty (who are essential to the equity and diversity missions of colleges and universities) view student affairs and diversity administrators as lacking academic credibility, students of color will remain ill served by retention efforts.

While the organizational chart varies at different institutions, I assume throughout this essay one of the most common arrangements. This arrangement entails an office of multicultural affairs or multicultural services, typically located under a vice provost or vice president (VP) for student affairs; usually separate from this office is the increasingly common vice president or vice provost for diversity or for equity and diversity, often reporting directly to the president or to the senior provost. Offices of multicultural affairs sometimes report to the VP for diversity, but both offices are typically outside of academic affairs and the research division. A nonrandom sampling of websites and organizational charts in June 2008, for example, revealed the following arrangements: Harvard University's Senior Vice Provost for Faculty Development and Diversity reports directly to the University Provost; the Associate Vice Provost for Institutional Equity of the University of Michigan, Ann Arbor, and its Associate Vice Provost for Academic Multicultural Initiatives both report to the Senior Vice Provost for Academic Affairs, although the college and school deans report directly to the Provost, rather than to Academic Affairs; the Director of the Multicultural Information Center at the University of Texas, Austin, reports to the Vice President for Student Affairs; Cornell University's Vice Provost for Diversity and Faculty Development reports directly to the Provost; the University of Virginia's Dean for African-American Affairs reports to the Vice President and Chief Student Affairs Officer and its Vice President and Chief Officer for Diversity and Equity reports directly to the

President; the University of Oregon's Vice Provost for Institutional Equity and Diversity reports to the Provost and is, in turn, the reporting unit for the Office of Multicultural Academic Support; Stanford University's Vice Provost for Faculty Development and Diversity reports to the Provost; Brown's Associate Provost and Director of Institutional Diversity reports to the academic Provost; Syracuse's Director of Multicultural Affairs reports to the Senior Vice President and Dean of Student Affairs; the Vice Chancellor for Equity and Inclusion of the University of California, Berkeley, reports directly to the University Chancellor, and its Associate Vice Provost for Faculty Equity reports to the Executive Vice Chancellor and Provost. Neither Texas nor Syracuse has Vice Provost- or Vice President-level positions in charge of diversity and/or equity.

This list represents a wide range of leading institutions of research and higher education, and yet, reading the mission statements and web pages of chief diversity officers at Harvard, Michigan, Texas, Cornell, Virginia, Oregon, Stanford, Brown, and Syracuse turns up not a single use of the word "research." In my sample, only Berkeley distinguishes itself by mentioning research as part of the mission of *both* its Vice Chancellor for Equity and Inclusion *and* its Associate Vice Provost for Faculty Equity. In addition, the Vice Chancellor's website prominently features links to academic units researching different aspects of diversity. Berkeley's Associate Vice Provost for Faculty Equity lists as part of her mission "conducting cutting-edge research on faculty equity," while the Vice Chancellor lists as one of his responsibilities to "encourage ongoing research to understand the effectiveness of activities here and elsewhere." Offices at the other institutions on this list typically describe their mission as contributing to their institution's diversity goals and educational mission by eliminating prejudice and discrimination on campus and by creating a climate for recruitment and retention of students, staff, and faculty of color. However, they fail to conceive of themselves (at least within their mission statements) as central to the research mission of the university (either as producers or disseminators of research), except insofar as they describe diversity as a precondition for academic excellence.

I would like to take it as given that more cooperation and communication between academic researchers and multicultural affairs is a desirable goal. Given this goal, the question arises, "What are the optimal conditions for generating the most productive cooperation and communication across the research/student affairs divide?" Another delicate question follows, "Are offices of equity and diversity as they are currently structured on most campuses more likely to bridge or to deepen that divide?" To address these questions, I would like to make three proposals.

1. *The research mission of the University needs to be front and center in multicultural affairs.* One place to start would be to ensure that the staff of multicultural affairs offices and student affairs include active academic researchers. In order for this to happen, it is necessary for senior administrators to think of the positions as research positions in the hiring process, possibly including turning some positions in multicultural affairs into joint, tenure-track faculty appointments with research departments ranging from math and economics to English and Spanish. Of course, the tasks that multicultural affairs staff members perform include providing emotional support for students, and often students' families. Clearly, not all researchers are necessarily qualified for this kind of work. The goal, then, should not be to hire only active researchers for such positions, but to allow the intellectual mission of the university to take a central role in multicultural affairs work by involving at least some active researchers in the work of multicultural affairs. A position in multicultural affairs, however, would only be attractive to the best researchers if it included teaching and involvement in an academic department, commensurate salary, and the resources necessary to conduct research.

Another way to center the research mission of the university in multicultural affairs would be to involve students as researchers in campus climate surveys, rather than as simply objects of such surveys. Similarly, research internships, particularly in science, technology, engineering, and math (STEM) fields could figure prominently in the kinds of opportunities that an office of multicultural affairs affords students. Opportunities for undergraduate research would help students to gain confidence in quantitative and scientific research without the stigma of the "remedial" courses that are associated with multicultural affairs at some campuses. Additionally, "honest dialogues" about managing cultural difference in labs could ultimately prove much more important to the success of students of color than workshops on difference set within residence halls, and could also engage the interest of science faculty.

2. *Offices of multicultural affairs and diversity offices must take the lead in disseminating current scholarship on race and education to the faculty and to the administration.* Faculty members of color are often disappointed with how ignorant many senior administrators are regarding research on race and inequality. For example, researchers from Claude Steele to William Bowen have conclusively demonstrated

the inutility of the Scholastic Aptitude Test (SAT) for judging success in college and beyond, and yet most universities continue to promote it as an important factor in admissions decisions.[2] How can we expect provosts and presidents to be aware of the growing scholarship on race and higher education, however, if their own diversity officers and multicultural affairs staff are not able or interested enough to keep up on and engage in this research? Quite simply, the era when faculty and administrators talk about race without data needs to end. Given how increasingly overburdened many multicultural affairs offices are, enabling them to remain on top of current research, to contribute to it, and to disseminate it within their institutions might mean additional resources, administrative restructuring, and/or a closer convergence between research faculty and multicultural affairs staff. I do not know what the correct answer will be, and it likely will vary by institution, but every institution needs to consider seriously how it can make research more relevant to multicultural affairs and multicultural affairs more relevant to research.

3. *In some cases, equity and diversity offices might actually get in the way of building substantive links between research faculty and multicultural student affairs.* Equity and diversity offices certainly have the potential to create one more layer of bureaucracy between the two. Given the fact that these offices exist, however, institutions should strategize about how to ensure that they do not exist simply to generate another set of disconnected events and initiatives. Rather, they should function to facilitate contact, communication, and collaboration between faculty and students from underrepresented groups. White faculty and faculty in the natural sciences and engineering should be central to this project, both because those faculty are in some cases the least likely to be aware of research on race and inequality and because underrepresented students are most likely to feel personally and socially alienated in their classes. How to mitigate that alienation and how to mitigate inequalities in educational preparation that might result from racism and structural inequality should be priorities of faculty in STEM fields and of senior university administrators. Solutions need not only to take STEM research seriously, but also to take seriously social science research on inequality and social interaction in academic learning and research spaces. Diversity offices are particularly well situated to bring these researchers together to find the necessary solutions, but they need to have the will and the research legitimacy on their campuses to make it happen. They might take as one model among others Stanford University's Medical Youth Science Program (SMYSP). Begun in 1987, the SMYSP mentors low-income

and underrepresented high school students in ways to promote their college attendance and entry into health sciences fields. It has three main aspects. The largest brings high school students to Stanford for a science-based summer program. Another includes weekend workshops focusing on college applications and career counseling, and a third seeks to enrich high school science teaching, in part by training teachers and counselors. Initiatives like this would go much further toward advancing academic diversity goals than a hundred roundtables about stereotypes, prejudice, and "difference" in academic residence halls. Instead of perpetuating the belief that inequality results from individual feelings rather than structural inequality, programs such as SYMSP target the material disadvantage and lack of opportunity faced by communities of color across the nation.

Should Universities Pursue Diversity or Social Justice?

I would like to make a modest recommendation that universities question the utility of words like "multicultural" and "diversity" and consider replacing them with the goal of contributing to social justice. In the interest of full disclosure, I should note that I have made more than a few arguments in my time in favor of multiculturalism and diversity. Furthermore, I recognize that many different kinds of projects have been articulated under these rubrics, some of which I support and some of which I do not. Few academics would even agree on the definition of a loaded term such as "multiculturalism." Despite such caveats, I am convinced that institutions of higher education should seriously commit to a vision of the university as tied to questions of social justice.

What seems most vital to me at this moment in the history of higher education is that we avoid being sidetracked by justifications for diversity that appeal to diversity for its own sake. These both risk repeating the errors of *Brown v. Board of Education* (which assumed that racial equality would follow automatically from racial integration) and lead to a lack of focus regarding *what* diversity, *when*, and *why*.[3] Brown, Kurzweil, and Tobin remind us that equity, as well as excellence, has a long if inconsistent history as a central concern of higher education in the United States. They argue that the existence of this country's vast system of higher education rests at least in part on the recognition by many different people at many different times of the role education can play in making society more egalitarian and in furthering the ends of social justice.[4] We must try not to let diversity for its own sake get in the way of keeping these more important goals in sight. It therefore makes sense to me to link faculty and student diversity goals to social justice programs when feasible. Furthermore, I believe that those initiatives, like the Mellon Mays Undergraduate Fellowship (MMUF), that have switched from a strict consideration of race to a consideration of race as one among many factors actually have the potential to strengthen the ends of social justice and racial equality more so than initiatives based strictly on increasing racial diversity. However, it clearly depends on what those "many factors" entail, and I would argue that they should not simply include "a commitment to diversity," but rather a more robust "commitment to social justice."

Bowen, Kurzweil, and Tobin offer a detailed account of the restructuring of the Mellon Foundation's MMUF:

> For the program's first 15 years, only members of "underrepresented minority groups" were eligible to apply. In light of [*Grutter v. Bollinger et al.* and *Gratz v. Bollinger et al.*] and the evolving needs of higher education, the MMUF mission statement was broadened: "The fundamental objectives of the MMUF are to reduce, over time, the serious underrepresentation on faculties of individuals from certain minority groups, *as well as to address the attendant educational consequences of these disparities*" (our emphasis). Then, to ensure that the Foundation's intentions were understood, the criteria for program eligibility were made more inclusive.... The new selection criteria include race as one factor among others...and they also emphasize the importance of a "demonstrated commitment" to the fundamental purposes of the program.[5]

I would like to draw attention to the difference that the italicized "broadening" of the MMUF mission statement makes. It shifts the emphasis of the MMUF from a goal of pursuing diversity for its own sake to addressing the negative impact of the absence of particular kinds of diversity. In addition, the shift in the eligibility for the fellowship actually targets more precisely students who are committed to improving the representation of underrepresented groups on university faculties. The underlying commitment of the SYMSP to social justice is even more explicit. The SYMSP website states the Program's belief "that helping low-income and ethnically diverse students reach their own educational goals is ultimately the most effective way to help improve the health care services available to underserved and low-income communities." The ultimate goal of the SYMSP, therefore, is not simply to diversify the health sciences, but to improve healthcare for low-income communities and people of color—a fundamental commitment to social justice. The success of these two programs is undeniable. SYMSP reports that, for its summer program, 100 percent of its students have been from low-income backgrounds, 58 percent have been African American, Latina/o, or Native American, 82 percent have graduated from four-year colleges, and 50 percent attend, or have completed, medical school or graduate school.[6]

Following the lead of programs like these, I would like to suggest three proposals to enhance higher education diversity initiatives.

1. "Diversity" and "multicultural" requirements in undergraduate curricula should substantively address the nature of structural inequality, racism, power, and privilege, rather than emphasize cultural diversity and tolerance of difference. Most universities introduced requirements in the 1980s or 1990s for a course or sequence of courses addressing multiculturalism and/or diversity. In many cases, the original intent of the advocates of these requirements was that courses satisfying them would substantively address the causes and consequences of structural inequality in U.S. society. However, on most campuses, a number of compromises were necessary in order to overcome faculty resistance. It is time that colleges and universities revisit those compromises. One common source of resistance was the belief that there were too few faculty members on campuses to staff enough courses unless a broader range of courses on cultural difference and social diversity were included. As a result of transformations across the disciplines in the past two decades, this objection no longer holds at many institutions.

Particularly with regard to students of color, I suspect that diversity classes that are inattentive to structural inequality and legacies of systematic racism can in fact impair retention efforts by leaving students feeling even more alienated from their faculty and peers. As an undergraduate at a not-so-diverse campus, I found encouragement by finding space in the curriculum where faculty and students addressed race, ethnicity, sexuality, and social justice in a substantive manner. I did not receive this from a "diversity" or "multicultural" requirement because, when I was in college, the university did not have such a requirement. Instead, I was able to take courses at all levels of the curriculum (often, the only course offered during a given semester) where the view of the world from a race-conscious perspective either was a starting point or was at least a subject of serious inquiry. This was not all that I took, of course. I also took (and loved) courses on Milton, Dante, European Renaissance art, modern Japanese literature, and chemistry. However, I could study these topics while also knowing that the reality and struggles of people like me were also worth studying at the university and that there were those at the university who considered it important to work toward a more egalitarian society. I could see that there were faculty and students who recognized the world as a place where racial injustice existed and structured both our society and the educational system. Knowing that I was on campus with others who shared a commitment to racial justice sustained me perhaps more than anything else did. For students of all backgrounds, understanding the nature of social inequality and how power and privilege function will contribute far more to the building of a better society than being introduced to "cultural competence" and the celebration of difference.

2. *Colleges and universities would benefit from restructuring equity and diversity offices into offices with a mission to enhance social justice.* Doing so would, to be sure, encounter significant criticism. One of the greatest obstacles, I believe, would come from development officers' and trustees' fears about the reactions of donors. I think, however, that in most cases such fears are unfounded and express rather the timidity of many development offices. I am not calling for academic institutions to have social justice as their *only* mission, only to consider it as the more substantive goal underlying their already-affirmed commitments to diversity. Exemplary cases of how a broad commitment to social justice can be integrated into a university's mission include efforts by Syracuse University and Brown University.

In 2005, Syracuse University unveiled a set of initiatives intended to bring that university closer to the surrounding indigenous communities, on whose ancestral land the university was founded. In addition to enhancing resources in Native American Studies and creating a student-learning center, the centerpiece of the new initiative is the Haudenosaunee Promise Scholarship Program. Providing full financial assistance to students who are citizens of one of the six Haudenosaunee nations (either in the United States or Canada), the Haudenosaunee Promise increased the number of incoming Native American students at Syracuse by 800 percent in one year, to 44 (only 30 of whom were funded by the program).[7] By seeking to address a legacy of economic disadvantage and educational discrimination, the university also met a significant diversity goal. Notably, the university doubled the number of incoming Native American students—even discounting those funded by the new program. The justification for the program, however, is not to increase diversity at Syracuse, but rather to make the university accessible to qualified students who might otherwise not be able to attend.[8]

Brown University's creation in 2003 of the Steering Committee on Slavery and Justice has received much more national publicity than the efforts of Syracuse to cultivate relationships with neighboring Native American communities. More striking than the actual report of the Committee, however, is the university's 12-point response. The response included a commitment to work with the City of Providence and the State of Rhode Island to "develop ideas for how the history of slavery and the slave trade in Rhode Island may gain its appropriate and permanent place in the public historical record." It also promised to expand a program to provide technical assistance to historically Black colleges and universities (HBCUs) and to engage in a wide array of endeavors to support and enhance Providence and Rhode Island public schools.[9]

I believe that more of these sorts of initiatives could arise if colleges and universities were to replace the secondary goal of diversity with the primary goal of social justice. Just consider the differences that a search might yield if advertised as a search for a Vice Provost for Social Justice, rather than a Vice Provost for Diversity. How much more broadly might the successful candidate be invited to envision her or his charge in relation to the campus and to local, regional, national, and global communities?

3. *Colleges and universities cannot shy away from measures of assessment and accountability in enacting their commitments to diversity and social justice.* This is a point that Bowen, Kurzweil, and Tobin have already made, but it bears repeating.[10] Institutions that make a public commitment to social justice should be able to demonstrate how they have measurably influenced social inequality locally, nationally, regionally, and/or globally. Have their efforts to improve the preparation of disadvantaged high school students and to make higher education accessible to them improved representation of students from underrepresented groups not only in college generally but also in STEM fields? How many students actively participate in endeavors related to social justice, sustainability, and/or community service after graduation? (This last question should be as important as how many go on to receive postgraduate education.)

How Can Student Activism Contribute to the Mission of the University?

In concluding this essay, I should note that part of my own success as a minority undergraduate, and later graduate, student certainly came from an active and effective office of multicultural affairs and from the presence of faculty members who demonstrated a visible commitment to racial equality and social justice in both their teaching and their support of students. However, another important element was a vibrant, political, and multiracial cohort of friends.

Active support of student initiatives around identity issues needs to be a key element in a broad-based retention program. Being involved, in my case, in a Latina/o student group, in a gay and lesbian student group, and a peace-and-justice residence hall enabled me to feel that, even if there were very few students like me on campus, I at least had a place (or places) where I could feel supported and affirmed. These were not always ideal spaces, of course, but they helped me immensely. One way that universities can foster such spaces is through the support of themed program houses, including racially themed residence halls. Philosopher Amie Macdonald has offered a robust argument in favor of these units.[11] Other ways are to encourage an environment in which students feel empowered to discuss their ideas with faculty and senior administrators.

To illustrate my point, I would like to share two experiences I have had as a faculty member witnessing senior administrative responses to student activists. During my second year as a tenure-track faculty member, white students brutally attacked a Korean student, who suffered brain damage because of the assault. Students called on the administration to denounce the attack, but the president's office responded with a brief statement saying that it did not want to do anything that might jeopardize ongoing criminal investigations. Students were, understandably, outraged that the administration could not even issue a general condemnation of hate crimes, without specific references to any single example. Several student groups, including Asian and Asian American groups, Latina/o and African American student groups, and a number of social justice organizations mobilized and staged a march that culminated in a half-day occupation of the lower floors of the university's administration building. As part of the march, every participant carried a carnation (in Korea, flowers are traditionally presented to teachers on Teacher's Day). The carnations were to be left at the administration building in protest, although once the students arrived, they decided to wait until the university president agreed to talk with them. I observed the sit-in at the administration building, and was appalled when the university president finally made an appearance to hear the students' grievances. She descended from the upper floor, flanked by security guards, as if she feared bodily harm from the carnation-wielding students at her own institution. From a distance,

I could see the thinly veiled contempt she held for the students and she actually turned her back on the designated student leader when he asked her why the administration could not issue a broad statement against hate violence. She could not even bring herself to look him in the face, and I was ashamed that this was my university's administration.

In contrast, at my current university, students recently became active in asking the administration to departmentalize the Ethnic Studies program. Again, the sophistication of the students impressed me. Ascertaining that the college dean was a principal block to departmentalization, they decided to present her with a "bad apple" award, which consisted, rather unpleasantly, of an actual bad apple. To her credit, however, this dean, despite not changing her mind, remained in her office for well over an hour to discuss the matter with them. I am sure that it was unpleasant for her and that she had many other things to do, and she could have dismissed them as disrespectful rabble-rousers, but she did not. Equally impressive to me were the more senior administrators, who repeatedly took time out of their schedules to meet with students and to hear their point of view. This was not a question of ceding authority on intellectual matters to students, but of taking time to educate students about how universities function, what constraints administrators face, and what kinds of reasons they base decisions on. Rather than condemning student activists, the administration at this university actually invited them to participate as interlocutors about the future of higher education.

Making students feel validated in this way makes them more sophisticated activists and citizens. They are more likely to feel good about their alma mater after graduation and more likely to feel valued while they remain on campus. It is quite likely that students in my first example will retain as one of their defining memories of the institution, not the beating of a fellow student, but the university president's response to their plea to her to recognize their pain, anger, and grief. Students in the second example felt respected by the administration and will likely remember those experiences of empowerment and mutual respect as a defining feature of their undergraduate careers.

Conclusion

During my visit to my alma mater, I shared with the faculty in the department that had invited me an article by higher education scholar Anthony Lising Antonio. In it Antonio seeks to address not the difficulties of recruiting and retaining faculty of color, but what the value of having faculty of color is. Furthermore, he does not approach the topic from the perspective of diversity (providing role models and mentors for students of color or creating a more pluralistic academy). Instead, he asks the question, what unique value, if any, do faculty of color bring to the university's *scholarly* mission? Antonio finds that professors of color "are much more likely [as much as 30 percent more likely] than are white faculty to place high importance on the affective, moral, and civic development of students."[12] Furthermore, they "are 75% more likely than white faculty to pursue a position in the academy because they draw a connection between the professoriate and the ability to effect change in society."[13] Clarifying this last point, Antonio adds that although a majority of *both* white professors *and* professors of color "believe that *colleges* should generally be involved in solving the problems of society...faculty of color are more likely to take *personal* responsibility for applying their talents to the cause of social change."[14] Antonio's research suggests to me that the link between diversity and social justice is, indeed, alive and well for faculty of color. Furthermore, those institutions that pledge commitments to students' civic and moral development and to social change central to their missions will be more successful at recruiting and retaining faculty of color. In turn, those institutions that recruit and retain faculty of color will be most successful at pursuing missions related to students' civic and moral development and to social change.

Colleges and universities will undoubtedly face criticism for centering social justice in their missions, and not all would even do so. However, a social justice mission, within the sciences, math, engineering, arts, and business, as well as within the social sciences and humanities, is perfectly within the range of ends a university might legitimately debate and pursue. Some of the examples I have given show the range of ways institutions have pursued this endeavor. Most colleges and universities in the United States have already stated commitments to diversity. Broadening that commitment to acknowledge the reasons why diversity has become important to begin with can help them to link diversity goals with other goals ranging from sustainability, peace and international cooperation, social welfare and equality, to global health and the elimination of hunger and poverty. My contention is, further, that those institutions that take social justice seriously as central to their missions of producing and disseminating knowledge will also be those most successful in diversifying their faculty and student body. They will also be those that contribute the most to the elimination of "pipeline" limitations that result from legacies of discrimination and ongoing structural inequality.

Notes

I would like to thank Satya Mohanty, who insisted that I actually did have something important to contribute to this discussion. I would also like to acknowledge those senior administrators who have demonstrated to me that, with some vision, hard work, and perseverance, we can accomplish truly amazing things, especially Johnnella Butler, Nancy Cantor, Johnnetta Cole, Dan Little, and, at my own institution, Linda Brady and Russ Tomlin. I would also like to acknowledge the many student activists I have had the honor to know at Willamette, Cornell, and Binghamton Universities and at the University of Oregon.

1. Peter Schmidt, "Cold Reality Intrudes on Diversity Conference in Disney World," *Chronicle of Higher Education* (May 30, 2008).
2. See, among others, William G. Bowen, Martin A. Kurzweil, and Eugene M. Tobin, *Equity and Excellence in American Higher Education* (Charlottesville, VA: University of Virginia Press, 2005), pp. 79–87; Claude M. Steele and Joshua Aronson, "Stereotype Threat and the Intellectual Test Performance of African Americans," *Journal of Personality and Social Psychology* 69.5 (1995): 797–811; and Claude M. Steele, Steven J. Spencer, and Joshua Aronson, "Contending with Group Image: The Psychology of Stereotype and Social Identity Threat," *Advances in Experimental Social Psychology* 34 (2002): 379–440.
3. See Derrick Bell, *Silent Covenants: Brown V. Board of Education and the Unfulfilled Hopes for Racial Reform* (Oxford: Oxford University Press, 2004).
4. Bowen, Kurzweil, and Tobin, *Excellence*, esp. pp. 13–38.
5. Ibid., 154.
6. Stanford Medical Youth Science Program (SMYSP), Stanford University School of Medicine, "Mission," http://smysp.stanford.edu/about/ (accessed June 15, 2008) and "Evaluation Results," http://smysp.stanford.edu/evaluationResults/ (accessed June 15, 2008).
7. Sara Miller, "Haudenosaunee Promise Succeeds in Helping Native American Students Attend SU," *Syracuse University News* (August 22, 2006), http://sunews.syr.edu/story_details.cfm?id=3428.
8. "The Haudenosaunee Promise at Syracuse University," Syracuse University, http://financialaid.syr.edu/scholar-haudenosauneeflyer.htm (accessed June 15, 2008).
9. *Response of Brown University to the Report of the Steering Committee on Slavery and Justice*, Brown University, Providence, 2007. See also "Brown University Committee on Slavery and Justice," http:/www.brown.edu/Research/Slavery_Justice/.

10. Bowen, Kurzweil, and Tobin, *Excellence*, p. 151.
11. Amie A. Macdonald, "Racial Authenticity and White Separatism: The Future of Racial Program Housing on College Campuses," in *Reclaiming Identity: Realist Theory and the Predicament of Postmodernism*, ed. Paula M. L. Moya and Michael Hames-García (Berkeley, CA: University of California Press, 2000), pp. 205–225, esp. 207 and 217–219.
12. Anthony Lising Antonio, "Faculty of Color Reconsidered: Reassessing Contributions to Scholarship," *The Journal of Higher Education* 73.5 (2002): 582–602, 591.
13. Ibid., 593.
14. Ibid., 594; emphasis in the original.

4

EQUITY AND EXCELLENCE FROM THREE POINTS OF REFERENCE

Daniel Little*

This is a very important piece of work, at this time, in this country, and in the globalizing world. The authors have done a tremendous service in exploring in detail the empirical situation of the wide and complex reality of their subject. And they have brought powerful and illuminating "what-if" simulations to bear on university policies in relation to existing demographic and educational patterns. The book documents the history, rationale, and interdependency of the twin values of equity and excellence. It simulates the effects of a "thumb on the scale" for socioeconomic status (SES) applicants. It attempts to measure the interactions of SES and race and refutes the notion that efforts to erase the inequalities of access associated with the former will substantially remediate the disadvantages of race—they will not. The book is an exceptionally important contribution to our understanding of these critical issues, and one that can help policy makers and university leaders make wise choices in the conduct and future directions of higher education in America.

Why are these issues so important at the present time? For several reasons. For the individual, we aim to achieve greater justice and greater democratic equality of opportunity. For issues of racial justice, we must achieve a greater ability of American society to fulfill its

---

* **Daniel Little** has served as Chancellor of the University of Michigan-Dearborn since 2000, where he is also a professor of philosophy. Previously he served as Vice President for Academic Affairs at Bucknell University and as Associate Dean of Faculty at Colgate University. He is a philosopher of social science, with a continuing interest in the foundations of sociology.

obligations of equality of opportunity and social mobility to African Americans. For the economy and society, we need a greater capacity to optimize the pool of talent to stop the waste of talent needed so badly in our society and economy. And for the strands of civility, we aspire to a greater sense of legitimacy in a society where everyone has the ability to realize her human talents through effective education. The authors rightly give special importance to the structural disadvantages associated with race in American society—a point I would like to emphasize as well.

I contribute here by bringing perspective from three places to which I have a lived personal connection: Detroit, the most racially segregated metropolitan area in the United States; the University of Michigan-Dearborn, a high-quality regional comprehensive university; and the People's Republic of China, where higher education and the issues of equity and excellence are now at center stage. Each of these places has an unexpectedly tight relationship to the intertwined social values of equity and excellence in higher education.

Detroit

Living in the Detroit metropolitan area, it is impossible not to recognize the enduring, persistent, and intractable effects of racism and its legacy in America. And similar findings would be true in many other U.S. cities: Chicago, New York, Cleveland, Oakland, or Peoria. These are the problems of urban and metropolitan America. The degree of residential segregation in the Detroit metropolitan region of about 5 million people is extreme. The inequalities across race of basic components of human well-being are striking and persistent: income inequalities, health disparities, infant mortality differences between white and black families, levels of employment, and—most important for our discussions here—levels of educational attainment. Whether we measure inputs—income levels, quality of schools, availability of healthcare resources—or outputs—levels of employment, wage levels, health status, and educational attainment—we find gross and racially specific differences across white and black populations in the Detroit metropolitan region.

First, some of the basic facts. Detroit represents one of the most racially segregated metropolitan regions along white-black lines in the country. The 2000 census demonstrates that Detroit represents the second highest "dissimilarity index" among U.S. metropolitan areas (after Gary, Indiana, and ahead of Milwaukee, New York, and Chicago). Racial inequality and its history are written into the geography of the Detroit metropolitan area—from Eight Mile Road to East Detroit. (Tom Sugrue's excellent book, *The Origins of the Urban Crisis*, gives a fine-grained understanding of how this race-defined industrial landscape came about.) These racial barriers correspond to persistent poverty and employment disadvantage in the city. Seventy-five percent of children in Detroit Public Schools (DPS) are eligible for free or reduced-cost lunch; 21.7 percent of families were below the poverty line in 2000 (2000 census). Youth and adult unemployment are staggeringly high. Finally, health disparities along racial lines are persistent and debilitating. Infant mortality rates for black families in the city of Detroit are more than triple those of white families: 17/1000 compared to the white rate of 5/1000 in Michigan.

Of particular concern for this volume, educational disparities between the city and the suburbs are entrenched and long-standing. Detroit Public Schools serve an overwhelmingly black student population (reflecting the residential segregation of the city). The high school graduation rate for DPS was estimated at 44 percent in 2002–03 and 61 percent in 2003–04. Other estimates hover around 50 percent. This compares to 83 percent in Dearborn (a middle-income suburb) and 92 percent in Farmington (a high-income suburb). Racially and economically mixed suburban schools have a better educational record than Detroit, including Southfield (91 percent) and Westwood (86 percent) (www.michigan.gov/cepi). The overall picture in Detroit is one of poor schooling and unsatisfactory outcomes (with some very encouraging exceptions—Renaissance High School [94 percent] and Cass Tech [98 percent]). But the pyramid is narrow at the top and the number of well-prepared college-bound high school (HS) seniors is very low in the Detroit Public Schools.

Two points seem particularly important for our discussion here. First, racism itself—both subjective and structural—is the chief cause of these inequalities. This American city has been encapsulated within a set of barriers to social opportunity and social mobility that make it exceedingly difficult for young African American men and women to find their way out of the cycle of poverty and disadvantage. And second, educational opportunity is one of the most important factors in improving the life prospects of the people of Detroit. (Just imagine the thought experiment: every Detroit child is served by an effective, well-resourced school, with high levels of educational attainment, 90+ percent high school graduation rates, and substantial and widespread preparation for college. These patterns of racial disadvantage would presumably melt away!) Conversely, the lack of such opportunity is one of the most discouraging symptoms of this continuing social crisis in Detroit. (Significantly, issues of educational opportunity play a central role in the agenda of one of Detroit's premier racial justice organizations, New Detroit.)

I have used the word "crisis" several times. I do not think that Sugrue's use of the term "crisis" is rhetorical or "over the top." It is impossible to drive through Detroit without thinking, this is a social crisis of massive proportions; it will lead to worse problems in the future; and our society is doing very little that works effectively to turn the situation in a different direction. And the situation is worsened by the view that is sometimes taken within middle-class America: "Race is a twentieth-century problem," "We can now move beyond racially specific social and educational policies." And I do not believe that the issue of black urban racial disadvantage can be subsumed under more general topics such as "multicultural diversity" or the observation that "many racial and ethnic groups in America face obstacles that need to be addressed." The latter point is correct. But the issues that derive from black-white racism are specific and impactful problems with a specific American history, and they need to be addressed deliberately and specifically.

If we mentally review the dimensions of urban racial disadvantage—inferior healthcare, low income, poor schools, limited job opportunities, unsafe neighborhoods, and poor housing—it is perhaps not an absurd simplification to assert that jobs and education are the critical variables in this complex system of causation. And of the two, education is a necessary condition for the first. So dramatic improvements in the availability, quality, and effectiveness of public education at all levels is a key policy tool for ending the persistent patterns of racial difference in American society.

So improving the quality of schooling that is available to economically and racially disadvantaged people in Detroit is crucial; improving preparedness for postsecondary attendance is crucial; and creating opportunities for success within colleges, universities, and vocational training institutions is crucial. And universities can be part of the solution in each of these areas. We can encourage our schools of education to work constructively with urban and metropolitan public schools. We can design and carry out middle school and high school "college preparation" programs that enhance skills and also enhance the student's emotional and cognitive orientation toward college attendance. We can design "bridge" programs that help students make the transition from relatively undemanding science and mathematics programs in underfunded schools to the demanding course work they will be expected to complete upon arriving at the university. And we can work with focus and intelligence at the task of creating campus environments where economically disadvantaged and racial groups will find their attendance supported and encouraged.

It is very important and interesting to observe that there are very big differences in schooling effectiveness and success in the Detroit Public Schools—which gives a basis for some degree of optimism that the involvement of schools of education in partnership with urban and metropolitan school systems can make a big difference. The story in metropolitan Detroit is not one of uniform failure; it is possible for urban high schools to succeed in educating the young people served to high standards. Out of the city's 19 high schools, several have graduation rates and ACT scores that are competitive with good high schools throughout the state. So Cass Tech and Detroit Renaissance high schools witness graduation rates in the high 90s, and ACT scores between 20 and 22. At the other end of the spectrum, we find high schools like Ford and Denby with graduation rates of about 50 percent and ACT scores of about 15.

So what are the consequences of these patterns of racial segregation and disadvantage? They are severe: persistent poverty and youth disaffection; profound diminishment of opportunity and human development; and consistently poor opportunities for jobs, education, and healthy lives. Things are not changing—or at least, they are changing only slowly and for a small subgroup of the Detroit African American population.

THE UNIVERSITY OF MICHIGAN-DEARBORN

Now I would like to shift gears to give some focus to the kind of institution that I represent, the high-quality regional comprehensive public university. UM-Dearborn is located on the edge of the city of Detroit, in the inner-belt suburban city of Dearborn, and in the center of the Detroit metropolitan area. The University of Michigan-Dearborn represents a real success story in American higher education, in my opinion, and one that has its counterparts throughout the country in the form of other high-quality, nonelite public universities. What I have observed at UM-Dearborn is a highly attractive combination of the two values we are discussing in this volume: excellence with equity. UM-Dearborn provides real excellence in undergraduate education. And this education is delivered successfully to low-SES students. Alums produce high achievement in business, government, and nonprofit leadership. The faculty are entirely comparable to the liberal arts counterpart institutions where I have worked and taught, with a good representation of research-active faculty and a very high percentage of effective and committed undergraduate teachers. The institution and the faculty are highly committed to undergraduate success. And the campus community is unified in a vision of the role of the university that emphasizes engagement and partnership with the metropolitan region that we serve. This is a quality university that succeeds in its most basic task: educating a diverse range of talented students to high levels of attainment. We are able to recruit highly qualified faculty; we are able to motivate them to give their best efforts in the classroom; and we are able to support them in their scholarly research with good results.

Some basic Dearborn SES facts are as follows: the campus serves 9,000 students, with about 6,500 undergraduates. Sixty-five percent of undergraduate students are first-time baccalaureate students; 11.2 percent come from families with family income below $25,000, and another 16.6 percent come from families whose income is between $25,000 and $50,000. Students from racial and ethnic minorities include African Americans (7 percent), Hispanics (3 percent), Asian Americans (7 percent), and Native Americans (1 percent); and, unusually in the United States, the UM-Dearborn population has a large number of Arab American students (perhaps 8–10 percent).

UM-Dearborn falls within a segment of American higher education that is critically important for the future. Consider the scale of public higher education. Roughly 6,110,000 students are served by public four-year institutions nationwide. The American Association of State Colleges and Universities (AASCU) represents the majority of the nonflagship institutions within this mix, with 3,450,000 students served by AASCU institutions (55 percent). (Compare this with the approximately 100,000 students served by the nation's top 50 liberal arts colleges.) These regional campuses are nonflagship state institutions serving undergraduate as well as masters and doctoral students. And I want to emphasize the point that this is an immensely important form of excellence and equity. There is a range among these institutions in terms of overall educational quality, resources, and outcomes, to be sure; but there are outstanding examples of AASCU institutions that embody this "high-high" combination: high excellence, high equity. These institutions create a set of opportunities that mean that students from a range of backgrounds, from middle class to disadvantaged, can get a high-quality undergraduate education for a total educational cost of about $8,000 per year, and can develop the preparation that will be needed for "next steps" in professional schools, graduate schools, and working careers. UM-Dearborn is a source of genuine opportunity for the students we serve, and it provides them with a high-quality and effective education.

What empirical data is available to allow us to assess the impact that public universities have on the first-generation students whom they educate? Paul Attewell and David Lavin undertake to do exactly that in *Passing the Torch: Does Higher Education for the Disadvantaged Pay Off across the Generations?*, published in 2007. Their research consists of a survey study of a cohort of poor women who were admitted to the City University of New York between 1970 and 1972 under an open-admissions policy. Thirty years later Attewell and Lavin surveyed a sample of the women in this group (about two thousand women), gathering data about their eventual educational attainment, their income, and the educational successes of their children. Analysis of their data permitted them to demonstrate that attenders were likely to enjoy higher income than nonattenders and were likely to have children who valued education at levels that were higher than the children of nonattenders.

The benefits of higher education in increasing personal income were significant; they find that in the population surveyed in 2000, the high school graduate earned $30,000, women with some college but no degree earned $35,000, women with the associate's degree earned $40,783, women with the bachelor's degree earned $42,063, and women with a postgraduate degree earned $54,545. In other words, there was a fairly regular progression in income associated with each further step in the higher education credential achieved. And they found—contrary to conservative critics of open-access programs in higher education—that these women demonstrated eventual completion rates that were substantially higher than four- to six-year graduation rates would indicate—over 70 percent earned some kind of degree (23). "Our long-range perspective shows that disadvantaged women ultimately complete college degrees in far greater numbers than scholars realize" (4).

So access to higher education works, according to the evidence uncovered in this study: increasing access to postsecondary education is the causal factor, and improved economic and educational outcomes are the effect. This is an important empirical study that sets out some of the facts that pertain to poverty and higher education. The study provides empirical confirmation for the idea that affordable and accessible mass education works: when programs are available that permit poor people to gain access to higher education, their future earnings and the future educational success of their children are both enhanced. It is a logical conclusion—but one that has been challenged by conservative critics such as Bill Bennett. And given the increasing financial stress that public universities are currently experiencing due to declining state support for higher education, it is very important for policy makers to have a clear understanding of the return that is likely on the investment in affordable access to higher education.

This makes the point about access; but what about excellence? If we could magically measure the overall quality of an average college graduate, incorporating achieved competences in imagination, problem-solving, quantitative skills, rigorous reasoning skills, communications skills, multicultural understanding, and ability to work well in teams, my perception is that we would not find a large gap in quality between the graduates of elite institutions and regional campuses such as Dearborn. And where there are perceptible differences in outcomes, it lies within our power to consistently work toward redressing these differences through quality improvement. At the moment I cannot document this with hard data; but it is a perception based on extended experience at UM-D and several elite liberal arts institutions. But the thesis is an important one: it asserts that American public education is working on both cylinders—access and excellence—and that it is not a major life disadvantage in intellectual development to a young person to attend the public regional university over the elite institution. Put it another way: the quality achievements in undergraduate education of a wide variety of U.S. universities and colleges are pretty narrowly bunched, without a large separation between elite and nonelite institutions. This should be a source of pride for all of us, because it is a central achievement for American democracy and the values of equality of opportunity.

Here I would like to emphasize a point about excellence that always needs focus. Excellence in a university is not guaranteed by the quality of the inputs—the credentials of the faculty, the quality of facilities, the amount of research funding. Rather, achieving a quality education is more like baking bread. The ingredients are the beginning. But constant attention to the processes is needed in order to keep the joint product working up to its maximum potential. If faculty lose focus on the importance of close intellectual relationships with their students; if they come to overvalue research time over classroom time; if deans and department chairs ignore signs of quality erosion; if faculty and leaders grow inattentive to important developments in pedagogy, curriculum, and content; and if university leaders fail to consistently emphasize the priority of effective teaching and learning—then high-quality ingredients will still lead to mediocre bread. Put it another way: there is an important intangible aspect of educational quality that is measured by academic values and shared commitment to students' learning that is a feature both of the people of a successful university and its institutional makeup. And institutions differ greatly in this dimension!

We should be clear about what the "elite" institutions add that is not provided by the regional public institutions. There are some differences that affect quality of undergraduate education that differ across the two categories of institution: student-faculty ratio, funded undergraduate research opportunities, and, of course, substantially deeper faculty research capacity. But I am suggesting here that these differences do not result in large differences in educational achievement. The elite institutions bring a potent mix of "social capital" to their students—a dense network of prestige, companies, helping alumni, and so on, which combine to create exceptional opportunities for the graduates of these elite institutions. So the personal advantage of attendance to the student has something to do with incremental educational excellence, but perhaps even more to do with social capital.

Given the value equation—access with excellence, equity with achievement—I describe here, it is highly disturbing to witness the precipitous decline in public funding for higher education. For example, the state of Michigan has dramatically reduced funding for its public universities over the past four years. Our campus is fairly typical in Michigan, and has experienced a 13.6 percent decrease in state funding in the past four years. Our current budget reflects just under 28 percent of state funding, with the rest represented by tuition revenue. (This percentage has changed from about 51 percent in 1990.) This means that tuition rates rise more steeply—or else the quality of the education we provide begins to slip. There is no third way of endless but harmless belt-tightening.

So my central point here is to give greater emphasis than perhaps Bowen et al. do to the democratic importance and the educational success of the high-quality regional state university. With talented faculty, committed leaders, and ambitious and talented students, and with reasonable support from public resources, these institutions can provide high-quality, challenging, and effective undergraduate education that permits their graduates to compete successfully in the most challenging environments—in the professions, in graduate schools, and in civic life.

Beijing

Let us turn finally to an international observation. In 2005 I had the remarkable opportunity of spending a month teaching philosophy of the social sciences to graduate and undergraduate students at Peking University (PKU). I met students from many parts of China. And I talked with presidents and other university officials in Beijing, Shanghai, and Suzhou. These issues of excellence and equity are of enormous importance for twenty-first-century China. Educational opportunities in China are highly unequal, especially across east-west and urban-rural divides. The "social capital" represented by professional status and family has great import for young people's outcomes. I had a young woman from Inner Mongolia who explained her background in these words:

> I was born in Inner Mongolia, in a small village located in the northwest part, and finished primary education there. When I was in my 12th, I left the village for a small town to continue my junior high school study, and I was lucky enough to be one of the few students who had successfully [*sic*] access to the only key high school in our city. At that time Peking University became my dream, and I had been struggling to be closer to my dream. However I failed for some reason at last. Now I am a student in Inner Mongolia University which, I should say, is a good university itself, we have diligent students and diligent teachers, however much less opportunity which might mostly due to the environment.
>
> Still, I think I am a lucky person compared with my classmates, some of whom even did not finish their middle school and most of whom did not have such good chance to study in Peking University, by the way, I was selected as an exchange student to study in Peking University last term. Although it was only for one semester, I achieved a lot. Such achievement was not limited to specialized knowledge but also English study. I had taken IELTS in May 14th, and got a 6.5. I was pleased even it was not a high mark, both for the efforts I had paid, and the most important, the improvement of my English I had made during the presentation.

And a young man from Xinjiang Vygur Autonomous Region described his route to PKU in these terms:

> He is a junior student of the college of life science in Peking University, majoring biology. Qitai, a county in Xinjiang Vygur Autonomous Region of the northwest China, is his hometown. It's famous in the Tang Dynasty for its key position in the north path of the Silk Road.

> Before attending the entrance examination for college, he has never reached any far places out of Xinjiang, although Xinjiang is the largest province which covers 1,600,000 square kilometers.
>
> Taking a glance at his book shelves, perhaps you'll figure out that his major fields are more likely humanities, because there are nearly books about philosophy and literature. Only the magazines referring to popular science could give some clues for his real majoring. Why does he become so? The key reason maybe that he's never had so many chances to contact with so many books before entering PKU. To tell the truth, he is a stranger to the humanities, just learning and reading all for interest. Concerning philosophy, he prefers to not only the Chinese philosophy like the Analects of Confucius, but also Western philosophy like Friedrich Nietzsche's power will theory. Poems is also his favourite section and Emily Dickinson is his fondest foreign poet.

The social outcomes and the differentiation that result from entry into one of China's elite universities are staggering. The great national universities such as Peking University, Fudan, Shanghai Jiao Tung, and Tsinghua appear to give their graduates a fast track into China's new economy to an extent that seems to dwarf comparable advantages within the U.S. economy conferred by attendance at an elite U.S. institution. And the degree of selectivity, and the transmission of past inequalities of access and resources through the admissions process, also seem to dwarf the counterpart processes in the United States.

China's admissions system uses a national college entrance exam. A common worry among Chinese academics and U.S. observers is the consequence of the importance of the entrance exams: that talented Chinese young people are pressured to shape their intellectual development for the test, emphasizing rote problems and memorization over a more open-ended capacity for imagination and problem-solving.

All observers appear to agree that China's education system is sharply stratified in terms of the educational opportunities provided to Chinese youth across the main dimensions of inequality: east versus west, urban versus rural, professional versus worker, registered versus unregistered city inhabitant. The cost of attending itself is a prohibitive obstacle—per capita farmer income may be in the range of $300–$400 per year, so educational expenses and fees of $1,000 per year are an unimaginable luxury. But likewise, the fees and expenses associated with elementary and high school attendance—let alone privately funded coaching programs—make educational differences at the earlier levels inevitable. Enormous differences in the quality of education offered in rural and western areas versus urban and eastern areas reduce the likelihood of a talented person from Gansu achieving a high university entrance score and gaining entrance to an elite university. It is not that these successes do not occur; rather, it is that the probability of success for two equally talented young people will be wildly different in the two social origins.

These inequalities are prevalent even if the examination system itself is a level playing field and a neutral "meritocracy." But there are variations in the admissions scores required by elite universities depending on the candidate's province; paradoxically, students from disadvantaged provinces sometimes need a higher score to gain admission to an elite university than their Shanghai counterparts (as a result of the effect of provincial quotas), as I was informed by a high official at an elite Shanghai university.

SES differences and regional differences have substantial impact on access and quality. School fees, family need for child labor, and new income opportunities associated with family labor all accumulate as factors deepening educational access inequalities.

The importance of family affluence and "social capital" in preparedness for the entrance examinations must be underlined. Children from professional families have much greater understanding of the process of admission; they have access to preparation courses for the entrance exams and the TOEFL; and, of course, they have the family resources necessary to pay the fees.

The disadvantages associated with poor rural schooling cascade to the university level; it is substantially more difficult for an ambitious student in rural areas to gain the knowledge and skills needed for admission to PKU or Shanghai Jiao Tung University. The best student I had in my course at PKU referred to the substandard English instruction she was able to receive in her village and the disadvantages this early schooling had created for her.

Well and good, we might say: the elite universities are tilted toward urban professional elites, while open to highly talented young people from poor and western provinces and regions. But at least the disadvantaged students can get a first-rate education at the provincial universities—University of Inner Mongolia, Gansu, or Sichuan. So we might hope that the differences in outcomes are less severe than the differences in access in the first place. But this appears not to be the case. China's leaders appear to have made the choice of favoring "elite" over "broad" quality in higher education. The Ministry of Education is placing great emphasis on achieving world-class universities in China—with no evident rhetoric of enhancing the quality of the more broadly available regional and provincial universities.

The consequences of these twin inequalities of access and quality in China include at least these. First, a massive waste of human talent. It is inescapable to conclude that there are vast numbers of talented Chinese young people who cannot gain admission to university or who gain admission to a notably inferior institution. So these young people will not develop their talents and capacities to the degree that they might; and this is a costly loss to China. Second, it is an antidemocratic injustice to the individuals who suffer this severe inequality of access and preparation. Their own lives will be less successful, less affluent, less satisfying, and less "upwardly mobile" than would be the case with greater educational excellence available to them. Third, but more speculatively, these regional and "class" differences are impossible to reconcile with the post-1949 emphasis on equality and a "pro-poor" orientation from the government and party. Sustained mismatch between declared values and observed outcomes can be the cause of deep social discontent—and there is evidence of this kind of discontent in the countryside in China today.

Are Chinese leaders and academics paying attention to these problems? Yes. The president of a regional Shanghai university said to me—an important social mission for her university is to bring students from the western provinces. But the evident disparities in quality of regional and provincial universities versus the elite universities are great and it is unclear that they will narrow anytime soon. China seems to be on a trajectory of narrow elite excellence rather than broad and accessible excellence.

An important obstacle to quality is the political and ministry domination of higher education. The party secretary is more powerful than the president or provost. Academic freedom is limited. The ability of faculty to determine the content of curricula is limited. And the bureaucratization of higher education is stultifying. This bears out the importance of one of Bowen's central assertions about the components of quality: independence, variety, and faculty governance.

Higher education is getting a lot of attention these days from the Chinese public and Chinese officials. There is great interest in China in forming partnerships with U.S. universities to enhance the quality of graduate programs. There is little evidence of concerted thinking about more effective undergraduate education. And there seems to be little prospect in the next decade of genuinely improving the probability of educational success of poor rural people in China. (One salient fact: nonregistered "guest" workers in Shanghai cannot send their children to public schools without a school fee—which of course they cannot afford.)

We might advocate a solution based on the experience of American public institutions: a national and adequately funded strategy of extending the scope and quality of regional universities. If China in 2020 could assure itself that one-third of rural youth could find a place in a good quality, effective regional university, the values of excellence and equity would be much closer to realization. And China would be enormously advantaged by the surge in human talent and knowledge that would be created.

THE UPSHOT

In each of the three areas of focus in these remarks, there is one recurring theme: high-quality regional universities are a critical part of the solution to the higher education needs of a country. These nonelite institutions should provide challenging, high-achievement education to the students they serve. And in fact they can do so, at reasonable cost. The "all-in" cost of educating an undergraduate student at a public nonflagship university in the United States is about $10,000 per student, including the state subsidy.

The democratic importance of high-quality public universities, including especially the good regional institutions, is unmistakable and weighty. So the decline of funding is deeply alarming. We cannot preserve the parity and balance of equity and excellence without adequate resources. One or the other must suffer. This places a great responsibility on leaders and faculty: to vigilantly preserve the excellence and impact of the education their institutions provide.

5

PRESTIGE AND QUALITY IN AMERICAN COLLEGES AND UNIVERSITIES

Steven J. Diner[*]

In the second half of the twentieth century, the United States developed one of the most comprehensive systems of higher education in the world. With community colleges, a myriad of public universities and colleges large and small, independent institutions of every size and mission, and research universities in both the public and private sectors, some form of higher education has been potentially available for at least a generation to all American high school graduates irrespective of their age, how well they did in high school, where they live, or their race, ethnicity, national origin, religion, or social class. Perhaps it is inevitable that such an inclusive system of higher education would also be highly stratified. But in the last two decades, the growth of popular, consumer-oriented rankings of top institutions, decreasing state and federal support for higher education, and rising tuitions have significantly increased social class stratification, as several recent reports have documented.

I view these developments primarily from the perspective of the institution I head, Rutgers-Newark. We are somewhat unique in the stratified world of higher education. Rutgers University operates as both a single institution and a state university system. The largest,

---

[*] **Steven J. Diner** has served as Chancellor of Rutgers University-Newark since 2002, and as Dean of Arts and Sciences from 1998 to 2002. He is also Professor of History. Before coming to Rutgers-Newark, Diner served as Vice Provost, Associate Senior Vice President and Professor of History at George Mason University and as a faculty member in Urban Studies and Chair of the Urban Studies Department at the University of the District of Columbia. He currently serves as President of the Coalition of Urban and Metropolitan Universities.

oldest, and best-known campus is in New Brunswick, but Rutgers also has campuses in Newark and Camden. The standards for appointment, tenure, and promotion of faculty are identical on all three of its campuses. On the other hand, admissions decisions are made locally, each campus has its own distinctive student culture, and prospective students can apply to each of the campuses.

Faculty and administrators on all three campuses take pride in Rutgers's stature as a major research university and its membership in the highly selective Association of American Universities (AAU). All faculty candidates for tenure and promotion, including new faculty appointments with tenure, are reviewed by a university-wide committee, where the same high research standards are applied to faculty from all three campuses. Rutgers-Newark has PhD programs in all of its professional schools except law (business, criminal justice, nursing, and public affairs and administration), in all of the sciences, and in select areas in the humanities and social sciences. We award approximately sixty-five PhDs a year.

Rutgers-Newark is one of the "streetcar universities" that emerged in the first part of the twentieth century to provide higher education to working-class commuter students in major cities. The private University of Newark, with schools of law, business, and liberal arts housed in a former brewery and a razor blade factory, was taken over by Rutgers University-New Brunswick in 1946, 10 years before the state assumed responsibility for funding Rutgers as the State University of New Jersey. It has a long tradition of educating first-generation college students, students from modest- or low-income backgrounds, immigrants, and children of immigrants. Although no longer entirely commuter, we remain deeply committed to our historic mission. Rutgers-Newark is one of a handful of American universities, mostly public urban research institutions, where students from modest- or low-income backgrounds are taught by top research faculty.

In accordance with our commitment to opportunity, we deemphasize SAT scores in our admissions process. The SAT is a class-biased instrument, and students who can afford expensive SAT-prep courses and even private SAT tutoring can raise their scores substantially. The correlation between SAT scores and academic success is questionable at best for all students. But it is a particularly poor predictor of success for those coming from the kinds of backgrounds as are most of our students. I recall one student telling me that he was 10 years old when he came to New Jersey from the Dominican Republic with his father. He knew no English when he arrived, so he did very poorly on the SAT, but Rutgers-Newark admitted him anyhow through our

Educational Opportunity Fund (EOF) program for low-income students. He graduated with a 3.95 GPA, went on to Rutgers-Newark's law school, and clerked for the chief justice of the New Jersey Supreme Court before joining a top New York law firm. Another student, who grew up in Jersey City, explained that she always wanted to attend Rutgers Nursing School but in several attempts "could not crack 1000 on the SAT." She too entered through the EOF program, and she graduated with a 4.0 GPA before taking an excellent job at a nearby hospital.

So I find it particularly troublesome that many institutions brag about the composite SAT scores of their entering class, citing it as a mark of the quality of their institutions. In addition to its class bias, the SAT cannot measure motivation, interpersonal abilities, artistic creativity, and much more. But more profoundly, the SAT was never intended to be a measure of the academic quality of an institution or its student body. The College Board initiated the SAT in 1926 at a time when many universities tried to limit the number of students who did not come from appropriate old-stock white Protestant social backgrounds, to *predict* the likelihood of academic success. Its proponents argued that it would help identify "diamonds in the rough," intellectually capable students from modest backgrounds who lacked the social polish colleges at the time considered so important—a goal it has never achieved. Its purpose today, as in the past, is to help evaluate the ability of a single student. The test was never meant to be a composite measure of the quality of colleges. It does not tell us anything about what students actually learn at an institution. But selective institutions across the country now routinely tout their average SAT scores as a mark of quality, thereby perpetuating the class privilege that the SAT fosters. The SAT is, of course, a significant component of the much discussed *U.S. News and World Report* rankings of the quality of colleges and universities. And since students and their parents pay close attention to these rankings, many colleges seek to raise their SAT profiles, further exacerbating the stratification of American higher education.

Like many urban and metropolitan universities, my institution admits nearly as many transfer students as first-year students. Most of the transfer students come from community colleges, and particularly from Essex County College, located two blocks from us, which enrolls largely students from Newark and other nearby low-income communities. We have strong articulation agreements and close working relationships with Essex and other community colleges, and transfer students from community colleges to our institution perform very well. Indeed, Rutgers-Newark students who have completed associate degrees at a community college have the same level of academic success as students who begin their college education at Rutgers-Newark. This is one more way in which urban universities such as ours work against the prevailing trend toward greater class and racial stratification in American higher education.

U.S. News and World Report ranks schools in other ways as well. For the last 13 years, it has ranked institutions by their racial diversity. In all 13 years, Rutgers-Newark has been ranked number one in diversity among national universities, defined as those institutions that award the PhD. This is a source of great pride to us, and we brag about it all the time. But in fact, it has at least as much to do with our location in one of the nation's major clusters of new immigrants as it does with our admissions policies and educational philosophy. Indeed, our students are much more diverse than can be revealed by the ratios of African Americans, Hispanics, Asian Americans, Native Americans, and whites calculated by *U.S. News*. Approximately 40 percent of our first-year students tell us that English is not their first language. Thirty-three percent were born abroad. Ten percent tell us their religion is Hindu, 10 percent Muslim, and 3 percent Eastern Orthodox. Thirty-seven percent come from families that earn less than $50,000 a year.

Leaders of American colleges and universities have argued for many years before the U.S. Supreme Court and elsewhere that diversity is crucial for the academic quality of the education we offer. Yet the diversity rankings of *U.S. News* are completely separate from its quality rankings. In other words, *U.S. News* treats diversity as a social descriptor of colleges and universities, but one entirely unrelated to academic quality. Needless to say, this is a sore point at my institution, which *U.S. News* places in the third tier of PhD-granting universities.

I would argue that in today's America and in today's world diversity is an integral dimension of academic quality. In my institution, we view our extraordinary diversity as a valuable resource for learning both inside and outside the classroom. Some faculty members have designed courses that specifically draw upon students' cultural and family backgrounds. One theater professor had all of his students produce oral histories of family members and friends who had migrated from abroad, and crafted these testimonies into a theater production on issues of culture and identity among immigrants entitled "Something to Declare." Another professor has students in a Portuguese Studies class interview residents of Newark's Ironbound neighborhood, looking at the often complicated relationships between the older Portuguese residents of the area and newer migrants from Brazil and Central America. Many of the students themselves are from Portuguese, Brazilian, or Central American backgrounds, thereby facilitating their access to neighborhood residents and giving them unique insights into the cultural dynamics. I have taught a seminar on public policy issues in America that arise out of racial, ethnic, and religious diversity. When discussing affirmative action, students talked freely about their own experiences, or lack thereof, with discrimination and ethnic stereotyping. When we debated bilingual education, many students talked about how they themselves learned English. In our global world and our global nation, serious exposure to cultural diversity is an essential element in quality higher education.

A growing number of scholars starting their academic careers today deeply understand this development. We have found that the extraordinary diversity of our student body is an enormous asset in attracting top faculty. Scholars in all fields get great satisfaction from teaching students with such a wide range of cultural backgrounds. Those in fields such as ethnic, urban, and global studies—areas of strategic focus on our campus and of great intellectual ferment generally—find the campus an incomparable place to work. We now award grants to faculty to develop or rework courses that take advantage of the unique learning resource that is our student body. And we conduct faculty professional development workshops on how to use that diversity to enhance learning.

Our students themselves tell us that campus diversity contributes significantly to their education. In our survey of graduating seniors, we ask whether they would recommend Rutgers-Newark to prospective students and if so, why. In the open-ended responses to this question, students repeatedly point to what they have learned from the cultural diversity of their fellow students. A few examples follow:

- "The campus is very multicultural and helped open my eyes to people I otherwise never would have been friends with."
- "Rutgers can make a person believe they are studying abroad because of all of the cultures and diversity encompassed on this campus. I feel as if I have traveled around the world at Rutgers-Newark."
- "I like the fact that I am going to school with different types of people. It allows for many different perspectives and good classroom discussions."

- "If high school students are interested in learning about different cultures, Rutgers-Newark is the school for them. This is the only school where I have learned from other cultures that are extremely different from mine."

A 2003 graduate, who has been highly successful at Merrill Lynch, spoke two years ago at our annual homecoming brunch. He attributed his success largely to what he learned as a student on our campus. "The Rutgers-Newark campus, indeed, is a microcosm of the global economy that enables us to interact with students from more than 150 countries and learn about their culture and tradition," he explained. "This has helped me tremendously in my job at Merrill Lynch. I am repeatedly called in to manage projects that are more global," he asserted, "as well as contribute to building my company's multicultural business."

Should not learning experiences like these factor into our assessment of educational quality? Should not our hierarchy of prestige place special value on the extent to which institutions graduate students from poor and modest circumstances? If we are serious about reversing the growing class and racial stratification of our colleges and universities, if we really mean what we say about the educational value of student diversity, then we must stop embracing the popular, simplistic, and elitist rankings of educational quality that so deeply influence American higher education today. Our nation surely will not diminish its growing class and racial stratification if it cannot reverse the stratification of our colleges and universities.

6

EMBRACING THE COMMITMENT TO ACCESS, DIVERSITY, AND EQUITY IN HIGHER EDUCATION

*Muriel A. Howard**

There has been much discussion over the last decade about how we, as leading institutions in higher education, are tasked with preparing the next generation of professionals to compete and succeed in a global economy. This economy, increasingly knowledge based, is reliant on the "product" we develop. If we liken the global economy to an engine, our product is the human energy that provides the fuel for that engine. Virtually every industry and commercial endeavor is reliant on the highest grade of fuel to maintain its performance. Academic institutions of higher learning represent the "refineries" in this process, taking the partially developed human energy that emerges from secondary institutions and converting it to the highest grade of fuel required by this global engine-economy.

Practically every nation of the world contributes some component to the global engine-economy. Institutions that embrace the world and invite its citizens through their doors are more apt and able to optimally refine the fuel needed by this engine-economy. These institutions produce a premium-grade, high-output source of energy.

---

* **Muriel A. Howard** is President of the American Association of State Colleges and Universities (AASCU), a post she has held since August 2009. Prior to assuming leadership of AASCU, Dr. Howard was President of Buffalo State College (State University of New York) from 1996 to 2009. Her professional and scholarly interests include support of education, educational leadership, representation of women and minorities, and public service. As AASCU President Dr. Howard advocates for public higher education at the national level, working to influence federal policy and serving as a resource for presidents and chancellors at 430 campuses nationwide.

When it comes to the question of diversity, institutions that go beyond simply teaching tolerance to embracing diversity position themselves more competitively. As world citizens, we must do more than tolerate differences in race and creed, ethnicity and geography. The global engine-economy demands that we do more than recognize and tolerate these differences. The performance of the global engine-economy requires that we actively engage our students to respect, promote, and celebrate these differences.

Enlightened individuals and institutions understand that there are unique, multicultural differences within and among people of the same race or nation. And these enlightened institutions value these differences. The most successful businesses are those that take strategic and outcome-focused steps to become informed about these differences. By doing so, they become suitably prepared for the interactions and interdependence that unfold during the planning, development, and implementation of a new project or venture. They also maintain their competitive edge, focusing on the corporate goal of becoming culturally literate to promote their business interests around the world.

The forward progress of the global engine-economy is impeded by ignorance. Ignorance, intolerance, and disrespect have an adverse effect on the workings of the global engine-economy. The role of higher education institutions has never been more important in the fight against ignorance and its ugly siblings: prejudice and exclusion. Exclusion, either by negligence or by design, is usually a prescription for disaster in most fields, but especially in ours, as the stewards of higher learning. Exclude a key line of code and a computer program is nothing but cyber-gibberish. Forget to cite a source and a book suffers from plagiarism; its author becomes the object of ridicule and censure. Leave out just one ingredient and even the best recipe can be ruined. Sure a cake can look like a cake, but without the sugar it will fail the taste test and fall short of its purpose.

Diversity takes effort, even when a college is located in an urban area. Just over a decade ago, the number of African American or other faculty of color who had served Buffalo State College for 25 years or longer could be counted on one hand. At the same time, only 16 percent of the college's students were from minority groups, while the city in which the college is located had a 40 percent minority population. This underrepresentation had three distinct yet interrelated elements: lack of access, lack of diversity, and lack of equity. The best approach to this problem required an action plan that addressed the three impediments that had kept Buffalo State College from embracing diversity, improving access, and achieving excellence.

At Buffalo State, we integrate inquiry with action. A multifaceted problem needs to be addressed with a multifaceted approach. Buffalo State College launched a comprehensive project designed to build on its institutional history. The project involved a broad representation of the campus community, ensuring an inclusive approach. This diverse group identified shared purposes and guided the eventual development of a new mission statement to appropriately reflect our core values. The Buffalo State College community is committed to the following: access to quality public education; quality teaching and learning; opportunities for individuals to realize their full potential; the rigors, joys, and fulfillment of intellectual discovery; supportive and collegial relationships; respect for diversity and individual differences; and service to society.

Let me take just one of these core values and embellish what is meant by "respecting diversity and individual differences." Buffalo State College wholeheartedly embraces the access and opportunity component of the mission statement of the American Association of State Colleges and Universities (AASCU): "We believe that the American promise should be real for all Americans, and that belief shapes our commitment to access, affordability, and educational opportunity and, in the process, strengthens American democracy for all citizens."

Another shared vision within AASCU is the ability to see the fate of a nation in the life story of a single student. The opportunity to exchange ideas, think strategically, and share intellectual viewpoints enables a college to improve a student's chances to succeed. The success of each of those students is inextricably tied to the success of our institutions, our communities, and our world. It is exciting to be part of AASCU, an organization that encourages its members to think beyond their campus boundaries to the world beyond. That is exactly where our graduates are headed—a highly diverse, competitive, and complex world. We need to help them get there, with the knowledge and preparation to be successful.

The world is powered by the global engine-economy discussed earlier. The processes we employ in refining the human fuel of this economy are critical to the progressive momentum of this engine-economy and its ability to power those nations not yet empowered with the means to lift their citizens out of lives of poverty and desperation. Is there a greater purpose on earth? Is there a nobler goal than to harness the power of the strongest nations in the world to maximize the potential of the weakest nations? Does this lofty goal bring more clarity to the mission of higher education? At Buffalo

State College, we believe it does. Given the task at hand, can our nation afford to leave "one mind behind" in its quest for economic justice at home or abroad? Is the inclusion of access, diversity, and equity in a college or university's mission statement really optional any more? We think not.

At Buffalo State College, our mission is expressed as follows: the goal of the college is to empower a diverse population of students to succeed as citizens of a challenging world. Recognizing the direction in which the knowledge based global engine-economy is taking the nations that have the resources to participate, the graduation of culturally intelligent professionals is no longer a luxury; it is a necessity. This intelligence will result in a competitive edge for our nation and for those institutions of higher learning that take advantage of multicultural opportunities for their students, faculty, and staff.

At Buffalo State College, we publish our policy on diversity and affirmative action at the beginning of each semester. It is not a trade secret. It is at the heart of our mission to educate future professionals to lead and to participate in their diverse communities. We selected a strong scholar and administrator to lead the equity and diversity efforts on campus. This person reports directly to the president, just as the vice presidents do for their respective administrative areas. We provide strong support for the work of the diversity office, and we outline equity and diversity expectations of all our line officers at the college. We have a number of courses that cut across disciplines incisively to meet our goal of embracing and celebrating diversity. In fact, our students must complete three credit hours of diversity classes in order to graduate. We devote residence hall space to learning communities, available to all student participants. We have launched a first-year learning community titled "And Justice for All," designed to help students understand and appreciate human diversity and the ongoing need for social justice. To ensure that the community's diversity is reflected in our student body, we provide scholarships to talented minority students and financial support to those who show great promise.

Buffalo State College made a conscious decision to celebrate its diversity by providing special recognition and support to the faculty, students, and staff members who promote diversity in their classes and out-of-class programs. We initiated the President's Award for Excellence in the Advancement of Equity and Campus Diversity and the Students' Awards (to faculty and staff) for Promotion and Respect for Diversity and Individual Differences. The Dr. Phillip Santa Maria Award for Student Leadership in Equity and Campus

Diversity recognizes student leadership in the name of a beloved dean who was a champion for social justice on campus and in the community and who recently passed away.

The Students' Awards nomination process, in particular, engages our students. It calls on them to identify and prepare an essay about the faculty, librarians, or professional staff members who have made special efforts to increase respect for diversity and individual differences. Through classroom discussions, curricular experiences, out-of-classroom activities, projects, discussions, or special mentor relationships, these individuals go beyond what is usually expected as a part of their regular job expectations or teaching assignments. The campus provides special financial incentives to students, faculty, and professional staff to support programs and projects that strengthen college diversity initiatives.

In the broader community, we are well known for collaborating with a number of professional, cultural, and religious organizations. For example, the campus is the official home of the Western New York Women's Hall of Fame. For the past five years, we have hosted the Conference for the Study of Peace and Understanding on October 24, which is United Nations Day. Community service is built into our mission statement, and service is valued with teaching and research in evaluating faculty.

Speaking of community, I want to describe the geographic environment in which Buffalo State College carries out its mission. Buffalo State is the largest and most comprehensive college in the State University of New York (SUNY) system, and the only SUNY comprehensive college in an urban setting. The City of Buffalo is the second-largest city in New York State. Our urban setting is among the college's assets in recruiting for diversity in all areas: students, faculty, and staff. We are situated in the most diverse community within the City of Buffalo, the West Side. Just five minutes from our campus is a Buffalo public high school that is attended by students who are learning English as their second language—Grover Cleveland High School.

In a true illustration of serendipity, this high school was actually the first home of the institution of higher learning that would evolve into Buffalo State College. Buffalo Normal School was the genesis of our college, and, given our mission of diversity and inclusiveness, it is fitting that it is now the home for high school students who speak more than 45 different languages and dialects at home. One could suggest that Buffalo State College need only open its doors to the community, and diversity is guaranteed. But it is not as simple as that, is it?

Access emanates from more than geography or the appeal of an urban setting and the amenities city life can offer. Access requires a rethinking on the part of families so they believe they have the right to a college education. They must believe that the pursuit and acquisition of a college degree can release a world of promise and a wealth of opportunities for their sons and daughters. Every family, immigrant or native born, has a rightful place at the table of higher learning.

Throughout our nation's history, the door of opportunity often had been locked to generations of Americans by discrimination, segregation, and exclusion. There are few ethnic or racial groups in America who have not been affected by this reality. Yet, once the door was opened, the responsibility fell to those who had been denied opportunity to walk through to the other side and take their rightful place in a nation moving toward integration and inclusiveness.

Whether a student comes to us from a family that never imagined itself as part of a college community, or from a nation where college opportunities did not abound, we welcome them. We provide all students with an education that enhances their intellect and enriches their life. We prepare graduates for productive, successful careers in a global society. We are positioned at the end of the educational pipeline and are the final stage of the refining process for this human energy to fuel the global engine-economy. The world looks to higher education institutions to ready future employees and entrepreneurs, thereby ensuring an adequate supply of vital human energy.

Buffalo State College also benefits from its proximate location to an international border. The United States and Canada have enjoyed nearly two hundred years of peace and prosperity. Canada is America's number one trading partner, with $1 billion in cross-border trade between our two nations taking place five minutes to the south of our campus at the Peace Bridge and twenty minutes to the north at the Lewiston-Queenston Bridge. Our national export trade to Canada exceeds the total for all of our other trading partners combined. An important component of the global engine-economy is energized and synergized in Western New York and Southern Ontario.

Canada also offers us a distinctly different approach to diversity and multiculturalism than what we have experienced in America. Canada, like us, is an industrialized, democratic nation that beckons immigrants from around the globe. The quality of life and standard of living in Canada is considered among the very highest in the world. The Canadian model of multiculturalism is often likened to a mosaic, as opposed to the American comparison to a melting pot. Multiculturalism in Canada is celebrated in very visible and tangible ways, thanks to robust funding from both the federal and provincial governments.

Members of this multicultural nation who emigrated to Canada generations ago were not systematically encouraged to lose their primary cultural identities—whether by changing their surnames, abandoning their native tongues, or downplaying their unique ethnic backgrounds. In some cases, one might characterize the experience of certain immigrant groups generations ago in the United States as one of suppression of their cultural identities. Could this be the reason why Canada has experienced far fewer incidences of racial tension that escalated into unrest and violence, or ethnic disputes that rose to that same level? One cannot say for sure, but it remains a possibility. Western New York was the final station on the Underground Railroad for many escaping slaves headed for freedom in Canada, their ultimate destination as they dreamed of unfettered lives for themselves and future generations.

Despite its historical role, Canada is not entirely free of some of the problems that have been gnawing at the United States for many years. However, the cultural climate in our neighbor to the north appears—at least to the casual observer—to foster greater understanding and appreciation of people from diverse backgrounds. The pieces of the mosaic are vibrant and strongly defined. They are separate and distinct, yet together compose a unified work of art, or in this case, a unified national identity. Immigrants continue to stream into both nations—the United States and Canada. We continue to make accommodations that facilitate their transition and eventual identity as citizens of our respective nations.

In Buffalo, the demand for immigrant resettlement, social work, and translation services has been continual since 1918, when the YWCA (Young Women's Christian Association) originally established the International Institute of Buffalo. The institute was initially formed to combat the exploitation of women who came to our nation and community to be the domestic help of the wealthy and elite. Although the original purpose of the organization is no longer relevant, it has been replaced by the multifaceted needs of a diverse group of immigrants that still requires services. The depth and breadth of the translation services offered by the institute provide a vivid illustration of the composition of the new immigrant population in Buffalo. Translation services cover 60 languages, from Albanian to Yoruba, from Bosnian to Vietnamese. Some of these languages were the first language of many of the students who now attend Grover Cleveland High School, mentioned earlier.

Let me now return to the discussion of the role of higher education in fueling the knowledge based economy. Situated in an urban community that—thanks to its geography—has been a hub of immigrant resettlement and opportunity, Buffalo State College has unique advantages in recruitment. However, these advantages are balanced by the challenges of accommodating such a disparate potential student body. These diverse individuals could very well compose a portion of the next freshman class. We at Buffalo State College have to be ready for this challenge, and we are preparing to do just that.

Embracing diversity and celebrating multiculturalism, while educating the future professionals needed to fuel the global engine-economy, must be fostered in a manner consistent with its importance. We strive to accomplish these two goals each and every day. We know we will not benefit from turning Buffalo State College into a melting pot that ignores differences. Instead, we look to the Canadian mosaic model as aligning better with our purpose.

Yet, I believe higher education can use another form of art as a metaphor for multiculturalism and the unique contributions each student can make to a vibrant, on-campus multicultural community. Weaving together threads of vastly different hues, textures, and sheens can produce a tapestry of exquisite quality. It forms a whole, to be sure—an artistic rendering that would be deeply compromised by the removal of even one thread. Upon closer look, the individual contribution of each thread is evident, just as the individual contribution of each student is apparent to the fabric of life on campus. The success of each student, indeed, is tied to the success of our institutions, our communities, and our world.

Let us look at the commitment to the expansion of access, the quest for equity, and the embrace of diversity not so much as a challenge, but as a wonderful opportunity to position higher education more strongly, more intelligently, in its role in the global engine-economy.

After all, the world is counting on us.

7

DIVERSITY AND EXCELLENCE IN AMERICAN HIGHER EDUCATION

*Eugene M. Tobin**

Shortly after the University of Virginia Press published *Equity and Excellence in American Higher Education*, Bill Bowen, Martin Kurzweil, and I participated in a panel discussion at the Brookings Institution. One of our colleagues on the panel, Amy Gutmann, President of the University of Pennsylvania, and a distinguished political philosopher who has written widely about education, democracy, and human rights, prefaced her remarks by referring to a *New Yorker* cartoon portraying a little boy tugging at Thomas Jefferson's coattails, looking up at Mr. Jefferson and saying: "If you hold these truths to be self-evident, then why do you keep harping on them so much?"

The reason we dwell on them, as President Gutmann observed, is that if we do not know the arguments in support of our most basic values and commitments—in this instance, diversity and opportunity in American higher education—these obligations and responsibilities will never become self-evident truths.[1]

In its earliest usages, the concept of diversity was rooted primarily in geography. When George Washington bequeathed funds for a national university, he did so in order that "the youth of fortune and talents from all parts" might, "by associating with each other...[be] enabled to free themselves in a proper degree" from "local prejudices and habitual jealousies."[2] Nineteenth-century educators and philosophers, as different as John Stuart Mill and John Henry Cardinal Newman, tended to think of diversity "in terms of differences in

---

* **Eugene M. Tobin**, a former President of Hamilton College, is Program Officer for higher education and the liberal arts colleges at The Andrew W. Mellon Foundation, and is the coauthor of *Equity and Excellence in American Higher Education*.

ideas," and both embraced "direct association among dissimilar people as essential to learning."[3]

Today, you would be hard pressed to find a college or university president who did not enthusiastically embrace the diversity of American higher education as one of the great strengths of our democratic society. "Diversity helps students confront perspectives other than their own and thus to think more rigorously and imaginatively; it helps students learn to relate better to people from different backgrounds, and to become citizens of the world."[4] Diversity in higher education also plays a crucial role in preparing students to be the leaders this country needs in business, law, and all other pursuits that affect the public interest.[5] Finally, as Patricia Gurin's research team demonstrated in their defense of the University of Michigan's use of race as one factor among many in the admissions process, "students who experienced the most racial and ethnic diversity in classroom settings and in informal interactions with peers showed the greatest engagement in active thinking processes, growth in intellectual engagement and motivation, and growth in academic skills."[6]

The editors of the *Economist* have observed that "the United States likes to think of itself as the very embodiment of meritocracy: a country where people are judged on their individual abilities rather than their family connections...." To be sure, the editors observe, "America has often betrayed its fine ideals...yet...today most Americans believe that their country does a reasonable job of providing opportunities for everybody.... But are they right?"[7]

The histories of America's oldest and most venerable colleges and universities are filled with iconic tales of efforts to admit students across a wide variety of barriers posed by race, gender, religion, ethnicity, and class. But the historical reality behind the struggles for diversity and inclusion is far more ambiguous and, in some instances, far less triumphant than we might like to believe. For much of the twentieth century, America's colleges and universities placed limited value on the importance of admitting students from different racial, religious, ethnic, and socioeconomic backgrounds. Indeed, many of the most selective institutions, fearful of alienating their traditional constituencies, consciously separated equity and excellence from "character" and "fitness."[8]

Today's barriers to higher education opportunity are subtler but just as intransigent. Although explicit policies to keep certain people out on the basis of race, gender, and religion have been eliminated, more "organic" barriers—such as poor academic and social preparedness, information deficits, and outright financial hardship—are limiting college opportunities for students from socioeconomically disadvantaged backgrounds, a group that contains more white students than minority students, even though racial minorities are disproportionately represented. These barriers are just as troublesome in their effects and in many ways more difficult to overcome than the explicit exclusion of individuals with unwanted characteristics.

One of the most interesting consensus outcomes to emerge from the extraordinary attention and controversy surrounding the University of Michigan affirmative action cases is a growing clarity that the goals and purposes espoused by our leading public and private colleges and universities are *public* purposes—that in Glenn Loury's words, "education is a special, deeply political, almost sacred civic activity." When you recognize that elite higher education is also a precious and scarce commodity where access to influence and power is rationed, how students are selected in this process is more than a ritual; it is, as Loury reminds us, a political act with moral overtones and, therefore, a matter of vital *public* interest.[9]

These concerns are heightened by the demographics of diversity and the demands of globalization. By 2028, according to the Business-Higher Education Forum, there will be 19 million more jobs than workers who are adequately prepared to fill them. Roughly 40 percent of the people available to take these jobs will be members of minority groups. And a large portion of these new jobs—especially jobs that offer competitive salaries and benefits—will demand skills and knowledge far beyond those of a high school graduate.[10]

What makes this look into the very near future so worrisome is the knowledge that the rate of increase of educational attainment in America has slowed to a virtual halt—right now, it stands at just above a high school diploma. And this has occurred at a time when the economic returns to a college degree are at an historic high: the market wants more educated workers, but a large segment of the college-age population is either unable or unwilling to respond.

The source of this problem is that college attendance rates over the past 30 years are stratified by parental income. And the gaps have barely narrowed. Students from the bottom income quartile (below $30,000) are about half as likely as their most privileged peers to enroll in a higher education program and are less likely still to graduate. This group contains more white students than minority students, even though racial minorities are disproportionately represented.

Our research suggests that the major reason for the gap in access is a lack of what we call "college preparation." This includes shortfalls in academic preparedness, but also the general inability of poor parents and underfunded schools to develop in their children and students the requisite practical knowledge, noncognitive skills, and motivation necessary to enroll in college, perform well, and graduate.

Young people whose parents' income is in the bottom quartile are half as likely to even take the SAT as those whose parents' income is in the top quartile. Our research (National Educational Longitudinal Study) indicates that the odds of taking the SAT and scoring over 1200—using the old scoring system with 1600 as the perfect score—are roughly *six times higher* for students from the top income quartile than for students from the bottom income quartile; and those odds are roughly *seven times higher* for students from the top income quartile than for students who are from the bottom income quartile *and* who are also the first in their families to attend college.

In the long run, the only way to solve the college preparation problem, or what we call the "supply-side block," is to attack it directly, through improvements in the schools that disadvantaged students attend, in the neighborhoods in which they live, and in the healthcare and other services that they desperately need. That will be a long and difficult process. Meanwhile, there is much more that can be done almost immediately at the undergraduate level to help a substantial number of deserving and qualified students from disadvantaged backgrounds.

There is no question that low-income students are underenrolling in the most selective, most expensive institutions, even after controlling for their college preparedness and other observable characteristics. But why? Is it a question of cost, academic preparation, and/or institutional policy?

Thanks to the cooperation of 19 highly selective colleges and universities, and to the assistance of the College Board, we now have at our disposal a rich new dataset that allows us to look "microscopically" at the more than 180,000 applications to these schools for places in the 1995 entering cohort, at the characteristics of those applicants offered admission, at the yields on those offers, and, finally, at the performance of the matriculants themselves as they moved through college to graduation.[11] Although these 19 schools are not a representative sample of the nation's 3,600 colleges and universities, as leaders of the country's higher education system, they exercise enormous influence and their policies shape the direction of higher education in the United States.

The president of one of these highly regarded colleges posed the central question to us early on in this project: "In applying to my university, is an applicant better off," he asked, "other things equal, being rich or poor?" Put another way: "Is there an admissions advantage associated with being poor, or with being the first member of your family to go to college, that is comparable to the advantage associated, let's say, with being a minority student, a legacy (the child of an alumnus), or a recruited athlete?"

The share of students at these schools from the bottom income quartile—and, as an alternative measure of low socioeconomic status (SES), those whose parents have no college experience—is unusually small. Students whose families are in the bottom income quartile represent 10–11 percent of all students at these 19 colleges and universities, while a little over 6 percent of these students are first-generation college-goers. Nationally, of course, the bottom income quartile is—by definition—25 percent of the population, while 38 percent of the national population of 16-year-olds have parents who never attended college.[12] When we combine the two measures of SES, and estimate the fraction of the enrollment at these schools that is made up of students who are both first-generation college-goers and from low-income families, we get a figure of about 3 percent.[13] Nationally, the share of the same-age population who fell into this category was around 19 percent in 1992, making this doubly disadvantaged group even more underrepresented than students with just one of the two characteristics.[14]

What is equally remarkable is that the percentages of students from low-income and first-generation college backgrounds at these 19 institutions do not change very much as one moves from the applicant pool to the group admitted, to those who enroll, and finally to those who graduate. This "equilibrium" is indicative of the fact that students with low socioeconomic status are treated the same in the admissions process as their peers are, and behave the same as their peers do from the time they are admitted until they graduate. This pattern suggests that once disadvantaged students make it into the credible applicant pool of one of these highly selective schools, they have essentially the same experiences as their more advantaged peers. In short, low socioeconomic status is not a factor in students' experiences as they move from applicant to matriculant to graduate.

What makes this finding of particular interest is the contrast it provides with admissions probabilities for three different types of applicants—recruited athletes, underrepresented minorities, and legacy students (children of alumni). Each of these groups has a better chance of being admitted than other applicants with the same test scores. Recruited athletes receive the biggest boost at these institutions, about 30 percentage points, followed by underrepresented minorities at 28 points, and legacies at about 20 points. In general, an applicant with an admissions probability of, say, 40 percent based on SAT scores and other variables would have an admissions probability of 70 percent if he or she were a recruited athlete, 68 percent if he or she were a member of an underrepresented minority group, and 60 percent if he or she were a legacy.

By contrast, students from low-income and low parental education backgrounds have an admission rate that is virtually identical to that of more privileged applicants with the same test scores, grades, race, and other characteristics. Our data demonstrate that students from low socioeconomic backgrounds are neither preferred nor at a disadvantage, compared to other students of the same race with similar scores in the admission process. Almost all of these institutions claim to be "need-blind" in considering applicants, and it appears that, in fact, they are.

Without going into detail, I can tell you that the pattern is the same for probability of enrollment, selection of major field of study, academic performance, and graduation. One of our major findings is that low-income and first-generation college students do not underperform academically—they receive exactly the grades we would expect them to based on other characteristics. Using survey data from earlier cohorts in our database, we were also able to determine that the pattern continues to later-life outcomes such as graduate degree attainment, earnings, career choices, and civic participation. In sum, once students from low SES and low parental education backgrounds enter the credible applicant pool, their socioeconomic status has very little bearing on their college experience or later life outcomes.

Given the dramatic underrepresentation of students from the bottom income quartile (11 percent) and first-generation college-goers (6 percent) at these 19 select institutions, two important questions emerge: first, is the current set of admissions preferences the best way to allocate scarce places at highly selective institutions? The second question is whether these institutions should continue to rely on a need-blind approach to admission.

Needless to say, the reasons for giving preferences to various groups differ radically, and we believe that some justifications are much more persuasive than others. Minority admissions preferences, as the Supreme Court noted in the University of Michigan affirmative action cases, serve educational and societal purposes—by enhancing the learning experience of all students, and by contributing to the civic, social, and economic life of the nation. Legacy/development preferences serve institutional purposes that benefit current and future students and faculty, and the preferences given to recruited athletes serve the purposes of the institution's athletic establishment and the interests of trustees and alumni/ae with strong feelings about athletics.

Our own biases lean toward admissions policies that strengthen what we call the complementarities between excellence and equity. Thus, among these privileged groups, we are most comfortable supporting minority preferences (because of the educational and societal goals to which diversity contributes), and least comfortable rationing highly desirable places at selective institutions to students whose athletic abilities serve narrower interests.

Now, what about the second question? Should colleges and universities continue to rely on a traditional need-blind approach to admissions?

Institutions who are interested (and able) to do more for potential students from low-income families are electing to provide even more generous financial aid to these students than they currently do already. As you know, a number of schools such as Harvard, Princeton, Yale, Brown, and the Universities of Virginia and North Carolina, have elected to replace loans with grants for students from lower-income families in an effort to increase enrollment of students from economically disadvantaged backgrounds; a number of other colleges and universities have since done the same. We hope these initiatives succeed but we suspect that giving a boost in admissions may be more effective in altering the socioeconomic composition of classes than increasing offers of financial aid to those who are admitted. The relatively high yields on current offers of admission to applicants from low-income families (higher than the average yields on all offers of admission) suggest that present-day financial aid policies are less of a problem than is reliance on need-blind admissions.

For institutions such as the 19 highly selective and wealthy colleges and universities in our study, we think consideration should be given to the alternative of putting "a thumb on the admission scale," and we are confident that the leaders of many of these colleges and universities would agree with us because several of them said as much in their amicus briefs to the Supreme Court in the Michigan cases.

A number of commentators have suggested that income- or class-based admissions preferences could, in fact, replace race-conscious admissions. In the simulations that we did for our book, low-income students are given the same admissions advantage as legacy students and underrepresented minority groups retain their current degree of admissions advantage. Under such a scenario, the admissions probability for low-income candidates at the schools in our study could be expected to increase substantially—from 32 percent at present to 47 percent. The admissions probability for all other applicants falls, as it would have to, but only from 39 percent to 38 percent. The explanation is, of course, the relative sizes of the applicant pools. Applying current group-by-SAT enrollment rates, our simulations show that the share of the class comprising students from low-income families could be expected to increase from 11 percent to about 17 percent. The minority share would (by assumption) stay constant at just over 13 percent, and the share of all other students would decline from 79 to 74 percent.

However, when we simulated the effects of *income-sensitive* admission preferences using the same preferences accorded to *legacy* candidates—*but this time eliminated race-sensitive admissions preferences*—we found that the share of students who are minorities fell by nearly half; although African Americans, Hispanics, and Native Americans are disproportionately represented among socioeconomically disadvantaged college applicants, the vast majority of disadvantaged candidates are white.

There are three kinds of potential "costs" associated with this hypothetical "thumb" on the scale: erosion of the academic profile, a decrease in alumni giving, and increased financial aid costs. For various reasons, only the last of these—increased financial aid costs—bears out substantially, and we estimate that such costs are probably affordable for the 30 or 40 wealthiest colleges and universities in the country.[15]

If enrolling a "critical mass" of minority students is a goal of higher education and society—as the overwhelming support for the University of Michigan from other universities and the corporate and military establishment in *Grutter* would seem to indicate—then income-based preferences are not a sufficient substitute for race-based preferences. But as my colleague, Bill Bowen, has observed, "Americans always seek the most painless alternative, and it's much easier for most people to be sympathetic to economic disadvantage than it is for them to understand and address the more deeply rooted and emotionally challenging issues of race in America."[16]

When the nation's elite institutions use affirmative action to ration access to their ranks, they "publicly confirm this ordering of moral priorities" and, as Glenn Loury reminds us, these decisions significantly influence the discourse on race and social justice in America.[17] Americans continue to disagree passionately over the short- and longer-term efficacy of affirmative action but, in general, most citizens see reversing the effects of an history of immoral race relations as a good thing and perpetuating those effects as a bad thing. But time is a real issue.

Indeed, the clock is already ticking on Justice Sandra Day O'Connor's widely quoted comment in the Grutter case: "We expect," the justice wrote, "that 25 years from now, the use of racial preferences will no longer be necessary to further the interest approved today."[18]

There has been much speculation about the basis and justification for Justice O'Connor's wishful and altruistic forecast. But if her 25-year expectation was intended as a firm road map, it is disconnected from any scholarly evidence based on black/white income convergence that suggests race-sensitive admissions will be unnecessary by 2028.[19] Given the uncertainty about the future composition of the Supreme Court, one would have to agree that the 25-year prediction is surely the "wake-up call"[20] that former Harvard president Derek Bok called it when he and Bill Bowen took a look back at their seminal 1998 study, *The Shape of the River*, to see how much work still needed to be done.

For everyone who cares deeply about ensuring the continued presence of racial diversity in higher education, and who is understandably concerned by the decade-long progression of antiaffirmative action legislation enacted to date in California, Washington, and Michigan, it may be instructive to remember what Bok and Bowen presented as the principal lessons learned from their research, the debates it stimulated, and some of the subsequent companion studies.

First, *the presumed educational benefits of diversity have been strongly affirmed*. Professor Patricia Gurin and her colleagues at the University of Michigan have documented a "consistent picture from both [their] research and the research of other scholars that shows a wide range of educational benefits when students interact and learn from each other across race and ethnicity."

Second, *race-sensitive admissions policies have increased substantially the number of well-prepared minority students who have gone on to assume positions of leadership in the professions, business, academia, the military, government, and every other sector of American life—thereby reducing somewhat the continuing disparity in access to power and opportunity that is related to race in America*. The evidence presented in the *River* shows that minority students admitted to academically selective colleges and universities as long ago as the mid-1970s have not only done well in the marketplace, but, most notably, have contributed in the civic arena out of all proportion to their numbers.

Third, *there is no evidence that, overall, race-sensitive admissions policies "harm the beneficiaries" by putting them in settings in which they are overmatched intellectually or "stigmatized" to the point that they would have been better off attending a less selective institution.* Indeed, the more selective the college they entered (holding their own SAT scores constant), the more likely they were to graduate and earn advanced degrees, the happier they said they were with their college experience, and the more successful they have been in their careers.[21]

Fourth, *progress has been made* in narrowing test-score gaps between minority students and other students, *but gaps remain—and so does the need for race-sensitive admissions policies.*

Overshadowing all the contentiousness and misunderstandings in the policy debates about race-sensitive admissions and class-based affirmative action is a harsher, more intractable reality. Today in America, we seem to lack the urgency, idealism, and confidence to tackle the sobering, unglamorous infrastructure problems of early childhood, primary, and secondary education. "What we're learning," as Kati Haycock, director of the Education Trust, has observed, "is that education is not like an inoculation, where if you do it once, you are set for life. It is more like nutrition, where you have to do it right and then keep doing it right."[22] Education at all levels—and especially at the precollege levels—needs and deserves a higher place on the nation's list of priorities.

If we take a variety of steps to encourage opportunity, support diversity, and demand excellence, perhaps the self-evident truths that we all need to keep harping about will truly become self-evident.

Notes

This essay is based on excerpts from *Equity and Excellence in American Higher Education* (Charlottesville, VA: University of Virginia Press, 2005), by William G. Bowen, Martin A. Kurzweil, and Eugene M. Tobin.

1. See The Brookings Institution, "Equity and Excellence in American Higher Education," transcript prepared from audiotape recordings, April 29, 2005, Miller Reporting Co., Inc., available at http://www.brook.edu/dybdocroot/comm/events/200050429.pdf, accessed July 15, 2005.
2. See Brief of Harvard University, Brown University, the University of Chicago, Dartmouth College, Duke University, the University of Pennsylvania, Princeton University, and Yale University as Amicus Curiae Supporting Respondents in *Grutter v. Bollinger, et al.* See also "The Will of George Washington," http://www.gw papers.Virginia.edu/documents/will/text.html, accessed July 14, 2005.

3. Neil L. Rudenstine, *"The President's Report 1993–1995"* (Cambridge, MA: Harvard University, 1995), pp. 4–5.
4. Brief of Harvard University, Brown University, The University of Chicago, Dartmouth College, Duke University, The University of Pennsylvania, Princeton University, and Yale University as Amici Curiae Supporting Respondents in *Grutter v. Bollinger, et al.*
5. Brief for Amici Curiae 65 leading American Businesses in Support of Respondents, in *Grutter v. Bollinger et al.*
6. Angelo N. Ancheta, "Briefing Paper, Revisiting Bakke and Diversity-Based Admissions: Constitutional Law, Social Science Research, and the University of Michigan Affirmative Action Cases," The Civil Rights Project (President and Fellows of Harvard College, 2003).
7. "Ever Higher, Ever Harder to Ascend," *Economist*, January 1–7, 2005, pp. 22–25.
8. See Jerome Karabel's magisterial study, *The Chosen: The Hidden History of Admission and Exclusion at Harvard, Yale, and Princeton* (Boston, MA: Houghton Mifflin, 2005) and Harold S. Wechsler's path-breaking monograph, *The Qualified Student: A History of Selective College Admissions in America* (New York: Wiley, 1977).
9. Glenn C. Loury, Foreword to William G. Bowen and Derek Bok, *The Shape of the River: Long-Term Consequences of Considering Race in College and University Admissions* (Princeton, NJ: Princeton University Press, 2000), pp. xxi–xxii and Loury, *The Anatomy of Racial Inequality* (Cambridge, MA: Harvard University Press, 2002), p. 132.
10. Business-Higher Education Forum, "Investing in People: Developing All of America's Talent on Campus and in the Workplace" (Washington, DC, 2002).
11. The schools in this special study include 5 Ivy League universities (Columbia, Harvard, Princeton, the University of Pennsylvania, and Yale), 10 academically selective liberal arts colleges (Barnard, Bowdoin, Macalester, Middlebury, Oberlin, Pomona, Smith, Swarthmore, Wellesley, and Williams), and 4 leading state universities (The Pennsylvania State University, the University of California-Los Angeles, the University of Illinois at Urbana/Champaign, and the University of Virginia).
12. See Steven Ruggles and Matthew Sobek et al., "Integrated Public Use Microdata Series: Version 3.0," *Minnesota Population Center Data Projects* (Minneapolis, MN: Historical Census Projects, University of Minnesota, 2003), http://www.ipums.org.
13. These estimates are consistent with the data in Anthony P. Carnevale and Stephen J. Rose, "Socioeconomic Status, Race/Ethnicity, and Selective College Admissions," in *America's Untapped Resource: Low-Income Students in Higher Education*, ed. Richard D. Kahlenberg (New York: Century Foundation Press, 2004), pp. 101–156.
14. The share of the national population of teenagers whose parents had not attended college and whose family income placed them in the bottom quartile in 1992 is determined by the authors' analyses using data drawn

from the National Center for Education Statistics' National Educational Longitudinal Study (NELS) of 1988 eighth graders.
15. These calculations were made based on the income distribution, net price by income, and list tuition of institutions in the Consortium on Financing Higher Education reported by Catharine Hill, Gordon Winston, and Stephanie Boyd in "Affordability: Family Incomes and Net Prices at Highly Selective Private Colleges and Universities," Williams College Project on the Economics of Higher Education, Discussion Paper no. 66, October 2003, pp. 5, 8.
16. William G. Bowen, "Extending Opportunity: 'What Is To Be Done,'" unpublished paper presented at the Macalester College and Spencer Foundation Forum, June 21, 2005, pp. 18–19. The published paper can be found in a very important anthology coedited by Michael S. McPherson and Morton Owen Schapiro, *College Access: Opportunity or Privilege?* (New York: The College Board, 2006).
17. Loury, *The Anatomy of Racial Inequality*, p. 138.
18. *Grutter*, 539 U.S. at 343.
19. See Alan Krueger, Jesse Rothstein, and Sarah Turner, "Race, Income and College in 25 Years: The Continuing Legacy of Segregation and Discrimination," Working Paper #9, Education Research Section, Princeton University, December 2004, accessible at http://www.ers.Princeton.edu/.
20. Derek Bok, "Closing the Nagging Gap in Minority Achievement," *Chronicle of Higher Education*, October 24, 2003, online edition.
21. In *Crossing the Finish Line*, Bowen, Chingos, and McPherson persuasively demonstrate that a large number of students from poor families and those who are African American or Hispanic—"undermatch." They study at less demanding four-year institutions (than they are qualified to attend), start at community colleges, or apply to no college at all. The authors found that students who "attend more selective institutions have a considerably higher probability of graduating than do comparable students who attend less selective universities." See William G. Bowen, Matthew M. Chingos, and Michael S. McPherson, "Helping Students Finish the 4-Year Run," *The Chronicle of Higher Education*, September 8, 2009, available at http://chronicle.com/article/Helping-Students-Finish-the/48329/ (accessed January 24, 2010).
22. See Karen W. Arenson, "Study of College Readiness Finds No Progress in Decade," *New York Times*, October 14, 2004, A26.

8

EDUCATIONAL INEQUALITY AND THREE WAYS TO ADDRESS IT[1]

*Michael S. McPherson and Matthew A. Smith**

There has recently been a heartening upsurge of interest in understanding and ameliorating the substantial inequality in college enrollment and completion between students from wealthier and poorer backgrounds in the United States. Attention has focused especially on the gap in attendance rates between families of varying economic backgrounds at the nation's leading colleges and universities.

Unfortunately such educational inequalities, understood as the correlations between socioeconomic status and educational outcomes, are also remarkably persistent: initial socioeconomic advantages allow parents to pass educational advantages to their children through unequal access to activities such as preschool programs, private tutoring, test preparation programs, private schools, prestigious (i.e. expensive) colleges, and so on. Such educational advantages are

---

*****Michael S. McPherson** is President of the Spencer Foundation. Prior to joining Spencer he served as President of Macalester College in St. Paul, Minnesota, for seven years. He is a nationally known economist whose expertise focuses on the interplay between education and economics. McPherson, who is coauthor and editor of several books, including *College Access: Opportunity or Privilege?*, *Keeping College Affordable*, and *Economic Analysis and Moral Philosophy*, was founding coeditor of the journal *Economics and Philosophy*.

Matthew A. Smith is a 2012 J.D. candidate at Yale Law School. He graduated Phi Beta Kappa with honors and with distinction from Stanford University in 2006, where he was awarded the Golden Medal for Excellence in Humanities and the Rheinlander Prize for Outstanding Work in Philosophy. He is a Point Foundation Scholar and worked for several years as a research associate at the Spencer Foundation in Chicago, Illinois. In that capacity he coauthored numerous scholarly works on ethical issues in higher education.

likely to translate into socioeconomic advantages, starting the process anew. Although this model is overly simple, it does highlight the fact that educational inequalities are deeply rooted in other social and economic factors, making them all that much more difficult and important to address. In any case, students from low-income families are far less likely than their higher-income peers to achieve the same levels of education. And we do mean *far* less likely.

If one classifies entering college freshmen by their family income, students from families earning more than $200,000 per year were more than four times more likely in 1999 to attend a highly selective[2] college or university than were those from families earning less than $20,000 per year (see table 8.1). It is hard to wrap one's mind around the fact that fully a quarter of these wealthy young college freshmen are at these top institutions, while just over 5 percent of poor kids who go to college are there. Allowing for the fact that people from poor families are significantly less likely than those from affluent families to attend college at all makes the situation even more extreme.

An illuminating view of the problem is provided by Bill Bowen and his colleagues in their recent book, *Equity and Excellence in American Higher Education*. Table 8.2, based on a similar table in their book, looks at the "pipeline" of young people from varied economic backgrounds. Getting to a top college requires graduating

Table 8.1 Distribution of first-time, full-time freshmen by income and institutional selectivity, fall 1999 (income in the thousands).

1999	<$20	$20–$30	$30–$60	$60–$100	$100–$200	>$200	All groups
Two-year public	39.0%	38.8%	35.5%	29.8%	16.8%	10.1%	30.9%
Two-year private	3.7%	2.7%	2.4%	1.8%	2.2%	3.5%	2.4%
Low select four-year	41.9%	41.8%	41.8%	42.4%	42.0%	35.4%	41.7%
Medium select four-year	9.7%	11.5%	14.9%	18.0%	23.1%	25.6%	16.5%
High select four-year	5.8%	5.2%	5.4%	8.0%	16.0%	25.5%	8.5%
	100%	100%	100%	100%	100%	100%	100%

Source: The American Freshman, 1999. Cooperative Institutional Research Program, University of California—Los Angeles, 1999.

Table 8.2 Percentage of the 1988 eighth-grade cohort who graduated from high school, took the SAT, and scored above 1200/1600 (by family income).

Family income	Percentage of cohort who graduated from high school (diploma or GED)	Percentage of high school graduates who took the SAT	Percentage of cohort who took the SAT	Percentage of those who took the SAT who scored above 1200	Percentage of cohort who scored above 1200
Bottom quartile	79.9	34.2	32.2	7.4	2.4
Second quartile	90.1	40.3	38.8	7.9	3.1
Third quartile	94.8	50.9	49.3	12	5.9
Top quartile	97.1	70.1	68.4	21.4	14.6

Source: The National Educational Longitudinal Study of 1988.

from high school, taking a standardized test such as the ACT or SAT, and doing well on it. Table 8.2 shows that the fate of less and more affluent students diverges at every step in the process, until it comes about that only 2.4 percent of students from the bottom income quartile score above 1200 on the SAT while 14.6 percent of students from the top income quartile do that well. It is thus immediately clear (as table 8.2 indicates) that differences in precollege academic preparation are a significant factor in producing inequality in attendance at selective colleges: among those who took the SAT, students from the top income group are about 36 percent more likely to score above 1200 than are those from the bottom group. Such differences in academic performance among students from differing economic backgrounds can be traced back to early ages, and it would not be reasonable to lay the bulk of the responsibility for those differences at the door of colleges.

At the same time, however, deficiencies in academic preparation are not the whole story in explaining differences in college access across income groups. Table 8.3 allows us to compare college access among freshmen by income group after controlling for differences in their academic preparation. Among the top third of test scorers in the high school class of 1992, 84 percent of students from the highest income quartile started a four-year college course immediately after high school, while only 68 percent of those from the lowest income quartile did so. Combining information on two-year and four-year colleges, we can see that 18 percent of top-scoring low-income students did not start college at all after finishing high school, compared

Table 8.3 Postsecondary enrollment rates of 1992 high school graduates by family income and math test scores.

Math test scores	Lowest income	Second quartile	Third quartile	Highest income
All institutions				
Lowest Third	48	50	64	73
Middle Third	67	75	83	89
Top third	82	90	95	96
Four-year institutions				
Lowest Third	15	14	21	27
Middle Third	33	37	47	59
Top third	68	69	78	84
Two-year institutions				
Lowest Third	33	36	43	46
Middle Third	34	38	36	30
Top third	14	21	17	12

Source: Sandy Baum and Kathleen Payea, *Education Pays 2004*, The College Board, p. 30

to just 4 percent of top-scoring high-income students who made that choice. To what degree these differences among well-qualified students are a simple matter of cash and to what degree they depend on subtler considerations of information, expectations, and motivation is unclear, but more than academic preparation is at work.

Given this stark picture, our goal in this essay is to think through the domains of educational inequality and, as importantly, emphasize that action in all of them is necessary to address educational inequality. Specifically, we argue that there are three domains, or groups of actors and the actions available to them, that deserve special attention. The first domain deals with colleges themselves: the financial aid policies at colleges and universities may have a substantial effect on the ability of students to attend. The most prominent example of this first domain is Harvard University's changes to its financial aid policies: in March of 2004 Harvard announced that "parents of families with incomes of less than $60,000 no longer will be asked to contribute to the cost of their child's education";[3] and even last year Harvard launched a "new initiative focuse[d] on ensuring greater affordability for middle- and upper-middle-income families through major enhancements to grant aid, the elimination of student loans, and the removal of home equity from financial aid calculations."[4] We will address this domain in detail later.

The second domain concerns financial aid policy makers. Examples of work in this area abound, but the most recent example is perhaps

Congress's attempt to curtail college prices. On February 7, 2008, the U.S. House passed a bill intended to put pressure on the nation's most expensive colleges and universities to lower their tuition costs and to expand substantially the maximum Pell Grant, a "floor" level of financing given to needy students.[5] As we argue, this domain is perhaps more important than the first in decreasing educational inequality in the nation as a whole. Policy makers can muster more resources and make far more dramatic changes than can college administrators, who must deal with the exigencies of running very costly institutions.

The third and final domain is perhaps the most important. Here the concern is not education policy itself but the broader social and economic forces that create vast disparities in precollegiate preparation and in turn structure educational inequality. Independent social science research across a wide variety of disciplines, including economics, neuroscience, and psychology, shows that educational disparities which manifest themselves in early grades (sometimes before preschool) are sufficiently durable to affect life prospects. As James Heckman puts it in his summary of the relevant research for *Science*, "[e]arly family environments are major predictors of cognitive and non-cognitive abilities. Research has documented the early (by ages 4 to 6) emergence and persistence of gaps in cognitive and non-cognitive skills."[6] These early childhood disparities, even when due not to lack of resources but to lack of cognitive and noncognitive stimulation, strongly correlate with economic status. By the age of six, students in higher income quartiles perform resoundingly better than their lower quartile peers on standardized tests such as the Peabody Individual Achievement Test in Math.[7] These disparities persist almost uniformly through elementary school. Even the best financial aid programs for higher education cannot be expected to overcome these initial disparities. In order to attenuate educational inequality, we must pay attention to the broader environment and its effect on educational inequalities.

Improving college opportunity for low-income students is plainly what Dennis Thompson has termed a problem of "many hands."[8] Notably, sustained progress is likely to require action in all three domains: individual colleges and universities, policy makers concerned with higher education, and policy makers who deal with poverty and education more generally. One danger with problems of this kind is that of "buck-passing" and "finger-pointing," as each actor tries to shift responsibility to someone else. Our purpose here is thus to stimulate work by actors in all three domains, each of whom can do something, and no one of whom can do everything. Or, as Theodore Roosevelt put it: "Do what you can, with what you have, where you are."

INDIVIDUAL COLLEGES AND UNIVERSITIES

It is unfortunately true that, despite the perception of many admissions officers at elite colleges and universities that they give some advantage to economically disadvantaged students in admission, a careful statistical analysis shows that this has not always been the case. Bowen, Kurzweil, and Tobin document the very low representation of economically disadvantaged students at a group of 19 selective colleges and universities as things stood in 1995.[9] One of the most striking findings from this research is that whereas athletes, racial minorities, and legacies (children of alumni) get a measurable advantage in admission, this is not the case either for lower-income or first-generation college students. After controlling for test scores, high school performance, race, and other measurable factors (other than socioeconomic status) there was no difference in the admission probability of more and less affluent students.

These findings from the 1990s have no doubt helped spark recent interest in the idea that selective and relatively affluent colleges and universities should do more to enroll low- and middle-income students (and, naturally, to help them persist and graduate) and we may hope that admissions practices at some places have changed as a result. Harvard's financial aid changes, discussed earlier, illustrate this concern; and to be sure, most of the universities and colleges involved in visible changes are wealthy private institutions. Among research universities, Yale has been quick to follow Harvard's lead and Stanford just eliminated tuition and room and board charges for students whose families make under $60,000 a year.[10] Among liberal arts colleges, Amherst and Williams College are just a few of the institutions that will have replaced many of their financial aid loans with grants beginning in the 2008–09 school year.[11] And some flagship public universities have instituted similar programs. The Carolina Covenant, for example, is a program instituted by the University of North Carolina at Chapel Hill that replaces loans with grants for families who make at or under 200 percent of the federal poverty level. The importance of these programs should not be underestimated. Prestige in many professional disciplines tracks degrees from elite schools. Almost three decades have passed in which either the president or vice president of the United States has had a degree from Yale. Similarly, Harvard, Stanford, and Yale are disproportionately represented among the alma maters of Supreme Court justices. Naturally this prestige bottleneck presents its own set of important normative issues: prima facie it seems morally problematic that a handful of institutions have a monopoly on access to the most prominent jobs. But so long as colleges continue to predicate admissions offers based on "merit," the bottleneck will likely remain and wealthy colleges and universities can and should use their policies to increase enrollment of low-income students.

The problem is that, acting alone, wealthy colleges and universities are likely to have at best marginal success in ameliorating educational inequality. The issue is one of scale. By this we mean two things. First, only elite colleges and universities have the resources to consider students on a "need-blind" basis. At the vast majority of U.S. institutions, offers of admission are predicated in part on a student's financial situation. Even some of the institutions listed by *U.S. News and World Report* as being among the top 20 national colleges are not need-blind. And some college guides go as far as to speculate that there are only 8–12 schools in the nation that are actually need-blind—that is, these schools do not just call themselves need-blind but actually keep admissions and financial aid entirely separate.[12] The fact is that most colleges depend heavily on tuition to finance their activities, and simply cannot afford to absorb the costs of enrolling too many needy students.

Second, the few elite colleges and universities that can afford to ignore students' financial position represent but a tiny slice of higher education. For example, in the fall of 2005, 17.5 million students were enrolled in an undergraduate (bachelor's or associate's) degree program,[13] while roughly 10,000 were enrolled as undergraduates at Harvard.[14] Were a change in financial aid policy to affect all students at Harvard, a total of 0.006 percent of undergraduates nationally would be affected. Given the preponderance of middle- and upper-income students at such elite institutions, the number actually affected by any change is likely to be substantially smaller. Moreover, there are reasons to think that the social effects of increasing the percentage of low-income students at elite institutions will be limited because almost all students who apply to and are competitive candidates for admission at elite schools also apply to and are admitted at safety schools (e.g. Syracuse is commonly seen as a safety school for Harvard applicants); an increase in the number of low-income students at elite schools will simply mean that the high-income students who would have attended those elite schools swap places with low-income students presently attending safety schools. This may be an improvement for both institutions—Harvard increases low-income enrollment and Syracuse gets more tuition revenue. But such a swap would affect such a narrow segment of higher education that, while of important symbolic and cultural value given the prestige elite schools command, it is not sufficient to address society-wide educational inequality in any fundamental way.

FEDERAL AND STATE FINANCIAL AID POLICIES

Of the three domains, general aid policies (state and federal) are by far the most promising in the near term if only for pragmatic reasons: of the three, they are likely to have the broadest impact with the least effort. Compared with the financial aid policies of individual universities, government policies are likely to affect a much larger fraction of the 17.5 million college students across the country. For example, in the school year 2003–04 around 62 percent of full-time undergraduate students received some sort of federal financial aid.[15] Whereas Harvard's financial aid policies will not even affect one-tenth of one percent of undergraduate students, the federal government's financial aid policies affect a majority. And, compared with the third domain (the broader social and economic factors that affect educational inequality—discussed later), federal and state financial aid present relatively simple policy options. It is much easier to change the amount of money the federal government allocates to students and the manner of that allocation than it is to change the underlying family and social dynamics that affect students from low-income families. This is not to minimize the importance of careful attention to the empirical effects and normative issues surrounding financial aid policies; but it is to say that this is one of the most straightforward ways to address educational inequality.

However, the news in this domain is not always heartening. First, it is worth emphasizing that, although it is difficult to generalize about the circumstances facing most undergraduates (precisely because their families' financial circumstances are so different), the American system of higher education finance throws up many roadblocks for students from disadvantaged backgrounds. Especially for parents who did not attend college, understanding what college will cost and how it can be paid for is a great challenge. Sources of funding are multiple and applications for aid are complex. Much financial assistance takes the form of loans and many disadvantaged families are deeply suspicious of the American credit system, and for good reasons. Many low-income students feel compelled to go to the cheapest available institution, whether or not it is educationally the best choice, and feel obliged to take on ambitious work schedules that interfere with progress toward the degree. More money is not a panacea for the educational challenges facing low-income families, but more money, better targeted and delivered more simply and reliably, is an essential part of any solution.

For all the attention that has been paid to "skyrocketing" tuition fees at American colleges and universities, too little attention has been paid to a primary cause of these tuition increases: a significant percentage reduction in the pecuniary support given by state governments to public education. Table 8.4 presents annual growth rates for spending on higher education (including that of families and governments) from 1980 to 2000. Although rates of growth have roughly paralleled GDP growth on a per capita basis, the aggregate rate of growth for public higher education spending fell substantially behind GDP growth during the 1990s. Two explanations for this trend are burgeoning Medicaid costs and growth in precollegiate education spending, both against a backdrop of resistance to tax increases—the former of which is likely to get worse given current and future demographic transition. States' investments in higher education simply have not grown fast enough to keep up with rising college populations. This needs to change, particularly because there is some evidence to suggest that investments in education are likely to yield more efficient long-term results than comparable investments in other fields.

Table 8.4 Comparative real growth rates in public higher education costs and GDP.

Period	Annual growth rates	
	Public higher education	GDP
A. Public higher education spending per student compared to GDP per capita		
1980–81 to 1990–91	1.92%	2.07%
1990–91 to 1999–2000	2.25%	2.13%
B. Aggregate public higher education spending compared to GDP		
1980–81 to 1990–91	3.14%	3.04%
1990–91 to 1999–2000	2.95%	3.38%

Note: "Education costs" are measured as education revenues minus revenues from auxiliary enterprises and student aid.

Sources: Educational costs and enrollments: Digest of Educational Statistics, 2002; GDP and Population: Economic Report of the President, 2003.

To cite one example, one of the premier questions in health economics is life expectancy: what explains variation in the average life span among different groups? A host of research indicates that education is an important factor. "The one social factor that researchers agree is consistently linked to longer lives in every country where it has been studied is education. It is more important than race; it obliterates any effects of income.... And, health economists say, those factors that are popularly believed to be crucial—money and health insurance, for example, pale in comparison."[16] There are a variety of possible explanations for this, not the least of which is that education gives people life prospects sufficient to motivate health-conscious behavior and the tools to discern health-conscious behavior. But the point is that education has impact well beyond its immediate economic ramifications and is not reducible to economic benefits, all of which makes the disparity in education between rich and poor all the more pressing.

Fortunately, not all the news is bad. In addition to Congress's recent interest, a number of authors, the present writers included,[17] have been pressing for serious reforms of the financial aid system. One of the simplest and most urgent changes needed is simplification: federal aid programs have proliferated and produced needless complexity. Similarly, deciphering and meeting the requirements of the Free Application for Federal Student Aid (FAFSA) is a challenge in its own right. Streamlining the process and making application easier could be expected to improve access for low-income students. Recent legislation has produced some progress in this direction, but much more can be done. Other more ambitious reform proposals deserve attention as well. Although the political will necessary for such reform may not currently exist, the hope is that when the will does exist the education community will be ready with deliberate recommendations. Greater attention to general financial aid policies is no doubt needed and, politically speaking, is the most likely to have some effect in addressing educational inequality.

Broader Social and Economic Factors

One way to explain educational inequality is to break it up into two interrelated but policy-wise independent factors. On this view, the relative dearth of low-income students in higher education would be explained both by the inability of low-income students to pay for higher education in a market economy and by their disproportionately poorer precollegiate preparation. While actions in the two domains discussed earlier (the policies of universities and colleges and financial aid policies more generally) may help mitigate the first factor, they are unlikely to have any effect on the second. Ameliorating educational inequality will require more than simply making college affordable—it will also require making college academically accessible to low-income students. This can be accomplished only by addressing the broader social factors, scholastic and nonscholastic, that contribute to disparities in academic preparation.

The task here is daunting in its complexity, implicating issues as diverse as neuroscience, pedagogy, and macroeconomics. For example, studies have shown that "poverty poisons the brain." Literally: "stress hormone levels tend to be higher in young children from poor families than in children growing up in middle-class and wealthy families...excessive levels of these hormones disrupt the formation of synaptic connections between cells in the developing brain—and even affect its blood supply."[18] The stresses brought on by poverty change the physical properties of the brain and affect (detrimentally) memory and language abilities. The upshot is that educational inequality is not merely culturally contingent; it has physiological components that sometimes impair the ability of low-income students to achieve at the same educational level as their upper-income peers. This suggests that addressing educational inequality is not merely a matter of making college affordable but of addressing these effects. If it is difficult to summon the political will to streamline the FAFSA, we can only imagine how difficult it may prove to carry out the intensive sorts of interventions that would help stem these neurological effects in the children of poor parents.

The bad news is that examples like these abound—the ecological effects of poverty are tremendous. The good news is that we (society collectively) can articulate many of these effects, a meaningful prerequisite for addressing them. As Nicholas Kristof put it in a summary of poverty for the *New York Review of Books*:

> we now have a much richer understanding of poverty than we did at the time of the launch of the War on Poverty [ca. 1964], and a much better hope of success if we try again. We know it's not just about more equitable distribution of assets, but that there are also crucial cultural issues to be addressed. We also have a better understanding of the tactics and policies that work, both in the U.S. and in poor countries abroad.[19]

It is important, at least theoretically, not to conflate educational inequality and poverty, although the two are inextricably intertwined.

Research makes clear that "disadvantage arises more from lack of cognitive and noncognitive stimulation given to young children than simply from the lack of financial resources."[20] In theory the same cognitive disadvantages that manifest themselves in children from low-income families would be replicated in upper-income families if children were exposed to the same stresses. In practice, however, this tends not to happen because upper-income families have the resources (time, money, etc.) to shield their children from harmful influences and mitigate incipient problems. For better or worse, the battle against educational inequality in this domain thus devolves into a proxy battle against poverty: if interventions are successful in attenuating the cognitive and noncognitive gaps between poor and wealthy students and in achieving parity in educational outcomes, the children of the poor may no longer be poor. "[E]ducation is a critical path out of poverty,"[21] which emphasizes our final point. It is unrealistic to expect educational inequality to disappear without massive attention to the underlying social and economic dynamics that fuel lower achievement for lower-income students. It is impossible to fully address educational inequality without addressing the inequalities from which it stems.

Conclusion

All of this is not meant to discourage. It is instead a call to action. Educational inequality does not just rest uneasily with the image of America as a "land of opportunity"; it transgresses what is widely understood as a fundamental concept of justice: careers and life positions must be open—and open not just in a formal but in a practical sense—to all regardless of where they started. This is not the case in America and the inequalities we observe suggest that the factors producing them must be deep-seated and many-sided. If we take seriously the goal of reducing these inequalities, we would do well to start from the assumption that successful action will require the efforts of many different actors. We have outlined three domains in which these inequalities need to be addressed. Because only elite colleges and universities have the resources to expand assistance to low-income students dramatically, institutional financial aid programs have significant social import but are unlikely to affect the fundamental disparities. In contrast, federal and state financial aid programs are essential in addressing college costs and are in principle more amenable to change. There is little excuse for not making college affordable for low-income students. Finally, any solution will require addressing the underlying social and economic factors that lead the children of low-income families to achieve less than their upper-income peers. Articulating these factors and calls to address them are important; what is left is to create the political will to actually do something about them.

Notes

1. Portions of this essay develop ideas originally published in the introduction to the book *College Access: Opportunity or Privilege?* ed. Michael S. McPherson and Morton Owen Schapiro (New York: The College Board, 2006). We thank Sandy Baum for helpful comments on an earlier draft.
2. The selectivity classifications in table 8.1 are provided by the Cooperative Institutional Research Program, and reflect differences in the average SAT scores in the institutions' entering classes.
3. Harvard College Admissions Office, "Harvard Student Recruitment Program," available at http://www.admissions.college.harvard.edu/prospective/hrp/index.html (accessed October 18, 2007).
4. *Harvard University Gazette*, "Harvard Announces Sweeping Middle-Income Initiative," available at http://www.hno.harvard.edu/gazette/2007/12.13/99-finaid.html (accessed February 13, 2008).
5. Jonathan D. Glater, "House Passes Bill Aimed at College Costs," *The New York Times*, February 8, 2006, available at http://www.nytimes.com/2008/02/08/education/08education.html?_r=1&scp=1&sq=education+pell+house&st=nyt&oref=login (accessed February 13, 2008).
6. James J. Heckman, "Skill Formation and the Economics of Investing in Disadvantaged Children," *Science* 312.5782 (June 30, 2006): 1900–1902.
7. Ibid., Figure 1.
8. Dennis F. Thompson, "The Moral Responsibility of Many Hands," in *Political Ethics and Public Office* (Cambridge, MA: Harvard University Press, 1987), pp. 40–65.
9. William G. Bowen, Martin A. Kurzweil, and Eugene M. Tobin, *Equity and Excellence in American Higher Education* (Charlottesville, VA: University of Virginia Press, 2005), pp. 101–105.
10. Yale University, Office of Public Affairs, January 2008, available at http://www.yale.edu/admit/freshmen/financial_aid/index.html (accessed February 13, 2008); Stanford University, Stanford Report, February 20, 2008, available at http://news-service.stanford.edu/news/2008/february20/finaid-022008.html (accessed February 21, 2008).
11. Amherst College, "Amherst College Will Replace Loans with Scholarships in Financial Aid Packages for All Students Beginning in 2008–09," July 19, 2007, available at https://cms.amherst.edu/news/

news_releases/2007/07_2007/node/14307/ (accessed February 13, 2008); Williams College, "Financial Aid," available at http://www.williams.edu/admission/finaid.php (accessed February 21, 2008).
12. Peterson's College Planner. Available at http://www.petersons.com/common/article.asp?id=3269&path=ug.pfs.advice&sponsor=1 (accessed February 15, 2008).
13. U.S. Department of Education National Center for Education Statistics (2006) *Mini-Digest of Education Statistics 2006*, available at http://nces.ed.gov/pubs2007/2007067.pdf (accessed October 18, 2007).
14. U.S. Department of Education National Center for Education Statistics, Institutional Information, available at http://nces.ed.gov/collegenavigator/?q=harvard&s=all&id=166027 (accessed February 14, 2008).
15. Note 11 earlier at p. 54.
16. Gina Kolata, "A Surprising Secret to a Long Life: Stay in School," *The New York Times*, January 3, 2007, available at http://www.nytimes.com/2007/01/03/health/03aging.html?_r=1&sq=education%20health%20care%20life%20expectancy&st=nyt&adxnnl=1&oref=login&scp=1&adxnnlx=1203443648-V1DQm3CNCrP6ToPM2djumw.
17. See, e.g., Matthew A. Smith, Sandy Baum, and Michael S. McPherson, "Financial Independence and Age: Distributive Justice in the Case of Adult Education," *Theory and Research in Education* 6.2 (July 2008): 131–152.
18. Clive Cookson, "Poverty Mars Formation of Infant Brains," *Financial Times*, February 16, 2008, available at http://www.ft.com/cms/s/0/62c45126-dc1f-11dc-bc82-0000779fd2ac.html?nclick_check=1 (accessed February 19, 2008).
19. Nicholas D. Kristof, "Wretched of the Earth," *The New York Review of Books* 54 (May 31, 2007).
20. Heckman, "Skill Formation and the Economics of Investing in Disadvantaged Children."
21. Kristof, "Wretched of the Earth."

9

CONSUMING DIVERSITY IN AMERICAN HIGHER EDUCATION

Gregory M. Anderson*

INTRODUCTION

In many ways the inauguration of President Barack Obama is indicative of the dualistic nature of race, opportunity, and democracy in American society. On the one hand, Obama's ascendancy to the very top of the U.S. power structure is a tribute to the American dream, a triumph of individual achievement and merit over the vestiges of group discrimination enshrined in a nation's shameful history of legalized slavery and segregation. Indeed, Barack Hussein Obama, the twenty-first-century version of Horatio Alger, has broken through a racial barrier many thought not possible to penetrate in their lifetime, to become the forty-fourth President of the United States. Yet, despite this individual accomplishment of historical proportions, the significance of race continues to restrict the life chances of millions of Americans. Indeed, whether focused on the so-called blackwhite achievement gap in education; the disproportionate number of African Americans and Hispanics incarcerated across the nation; the continued racial disparities in healthcare coverage, illness, and mortality rates; or the longstanding patterns of housing discrimination,

---

* **Gregory M. Anderson** is the Dean of the Morgridge College of Education at the University of Denver and a tenured Associate Professor in Education. Before coming to DU in 2009, Dr. Anderson was an Associate Professor at Columbia University's Teachers College, Program in Higher and Postsecondary Education. In 2006, Anderson was granted an extended leave from Teachers College to become the higher education policy program officer for the Ford Foundation in New York. He was responsible for overseeing one of the largest portfolios at the Foundation featuring both international and domestic higher education grants.

predatory lending practices, and labor market outcomes by race, there is sufficient evidence to remain skeptical about Obama's presidency signifying the birth of a postracial America.

If Obama's rise to power does, however, signify a radical break in history, it may be more accurate to view his presidency as ushering in the end of a Civil Rights narrative of racial redress born out of the turbulent 1960s and 1970s in America. To be clear, perhaps the ultimate irony of this sea-changing moment is that in our zest to finally be done with race and racism and to celebrate change and difference as essential to our humanity and democracy, a dangerous precedent has been set that privileges aspiration over reality. After all, when it comes to "ending" racism, whether in post-Apartheid South Africa or a postracial America, there is a tendency to look for shortcuts that sidestep more sober notions of what is possible when attempting to overcome centuries of violence, domination, misinformation, and fear of the "other." Rather than roll up our collective sleeves and accept the long, hard, and painful work of redressing the cumulative effects of racial and ethnic discrimination, it is much more palatable on the world's stage to put forth imagined communities of rainbow nations and melting pot societies.

To a degree this tendency to celebrate diversity is profoundly related to how we experience difference in an increasingly virtual world of unprecedented choices. Freed from the bounds of time and location, we can communicate across national boundaries and cultures in cyberspace, compare and consume an increasing array of goods and services, and witness events in high definition via blackberries, cellular phones, and flat-screen TVs. Diversity in higher education must therefore be interrogated in a context of unlimited access to information, entertainment, and media, in which being physically present is no longer a requirement to experience the human condition or to make split-second choices regarding wants and desires.

In an era of unbridled consumption and inequality is it so far-fetched to question whether diversity as a compelling societal interest has become a commodity in higher education, especially when taking into account that universities and colleges are situated at the cusp of opportunity, individual merit, competition, and social mobility? In exploring this question, I weave together two narratives. The first, more theoretical and historical in nature, talks about how race and diversity in U.S. higher education can be viewed as commodities in which difference is often heralded by universities and colleges but at the expense of failing to critically redress the present-day effects of racial and ethnic discrimination. The second narrative is more empirically driven and features a concise breakdown of descriptive statistics derived from the influential book, *The Shape of the River*, by William Bowen and Derek Bok. The underlying purpose of the Bowen and Bok data is to delineate the limits of the benefits of diversity in relation to small numbers of students of color from disadvantaged backgrounds gaining access to selective universities and colleges.

Diversity as a Commodity in Higher Education

Look no further than the brochures and catalogues of most universities and colleges and it is practically impossible to fail to appreciate the attention paid to minorities. Irrespective of the institutional type (selective versus nonselective, two- versus four-year, private versus public, etc.), higher education promotional materials are chock-full of images of minorities reading books under trees, writing intently during lectures, and walking the halls, with backpacks in tow. These images are emblematic of the times, in which diversity has become part of the higher education vernacular, a marker of something desirable, inevitable and beneficial. Before delving in greater detail into the relationship between diversity, race, and access to higher education, it is necessary to provide a working definition of what is meant by the term commodity.

For Marx, "a commodity is, in the first place, an object outside of us, a thing that by its properties satisfies human wants of some sort or another."[1] Marx went on to argue that a product becomes a commodity when it is "…transferred to another, whom it will serve as a use-value."[2] This process of transfer, according to Marx, contributes to fetishism, as the intrinsic value of a commodity based on human creativity, individual and collective effort, and a commitment of time becomes confused with its use-value to others. Furthermore, Marx believed that because commodities satisfied the wants or needs of others, people became increasingly alienated from their own humanity as they increasingly related to each other through the products they consumed. Finally, because enjoyment of commodities and the benefits (status, distinction, prestige, etc.) associated with ownership of products was determined by the amount of resources that an individual had at her/his disposal, the very act of consumption for Marx deepened stratification and inequality.

Simplifying and contextualizing Marx's concepts,[3] it is my contention that diversity has become a commodity in higher education, an alluring product dutifully showcased by elite universities and colleges as central to their missions but rarely resulting in significant numbers of students of color (especially from low-income and segregated backgrounds) gaining access. Thus, while the intrinsic value, to paraphrase Marx, of a credential earned from a selective institution of higher learning is first and foremost based on the return on the investment accrued by the individual, in the case of students of color, their visible presence on campus underscores an institutional asset or added value that is disproportionate to their low rates of participation. By heuristically evoking the concept of commodity, I am therefore referring to the immense use-value to others that a relatively small number of minority students attending selective universities and colleges provide in relation to the benefits of diversity enjoyed by these institutions and their predominantly white student bodies.

The benefits of diversity have been well documented over the years, ironically as a response to legal challenges to affirmative action and the use of race-conscious admissions policies in higher education.[4] Recent research has also chronicled the growing influence of multiculturalism, ethnic studies, and the internationalization of the American and Western higher education curriculum.[5] These changes in university and college curricula have been significant enough to warrant new categories being added to the most recent National Educational Longitudinal Survey.[6] For instance, selected additions to the taxonomy of courses from 1992 to 2000 include topics such as ethnic studies, multicultural education, immigration, and diversity in the classroom.[7]

It is not exaggeration to say that the discourse of diversity now informs almost every aspect of university and college life, whether in discussing learning outcomes and curricular content, outlining plans for future reforms and targeted goals to be reached, or defending the use of affirmative action in admissions at highly selective institutions. Moreover, in the aftermath of the Supreme Court affirmative action cases involving *Grutter* and *Gratz* versus the *University of Michigan*, even greater attention will be paid to documenting how higher education can meet the needs of an increasingly diverse society. There is little doubt that minorities have made significant and important gains in higher education, as the percentage of students of traditionally underserved or disadvantaged groups increased dramatically from 1980 onward. Between 1995 and 2005, for example, total minority enrollment grew from roughly 3.4 to 5 million, or by about 50 percent, and the minority share of the total enrollment rose from 24 to 29 percent during this same period.[8]

Yet upon inspection, it is critical to separate the demographic shift that has taken place across the nation from a more nuanced and balanced analysis of patterns of minority access and retention in higher education. Despite minority enrollment growth, African Americans and Hispanics received only 7 and 4 percent, respectively, of all baccalaureate degrees awarded in the United States according to 1993–94 figures.[9] If we fast forward a little over a decade to 2006, traditional college-aged African Americans (32 percent) and Hispanics and American Indians (25 percent) continue to lag far behind their white (44 percent) and Asian American (61 percent) counterparts in terms of rates of enrollment.[10] Looking at persistence rates also reveals a disturbing trend, especially with respect to Hispanics and African Americans. In 2003, African Americans had the lowest rate of persistence of any racial/ethnic group and by 2005 lagged well behind whites in terms of the total number of degrees (associate's and bachelor's) conferred. Similarly, Hispanics, despite an increase in the number of bachelor's degrees by 86 percent within a ten-year period between 1995 and 2005, nonetheless continued to lag behind all other racial/ethnic groups with respect to undergraduate degrees conferred in 2005.[11]

These statistics are even more alarming when factoring in the dampening effect of class or socioeconomic status on degree attainment. Regardless of race or ethnic background, young adults are eight times less likely to complete a bachelor's degree when their families fall into the bottom income bracket.[12] However, because African Americans and Hispanics are disproportionately (in relation to the percentage of the total group populations) from low-income or poor backgrounds, the cumulative disadvantages of segregation, depressed housing values, and high rates of unemployment have especially restricted their access to quality higher education opportunities.[13] As a consequence, it is crucial to consider simultaneously both race and class when attempting to separate the hype surrounding the benefits of diversity in higher education from the everyday realities and continuing constraints of racism in America. Unfortunately, the likelihood of implementing policies of redress has been severely diminished by the demise of the original intent and more expansive mandate of affirmative action. Moreover, to a large degree the waning of support for race-conscious policies over the last several decades can be directly linked to the concomitant rise of diversity as a hegemonic discourse in higher education.

THE SHIFT FROM REDRESS TO THE DISCOURSE OF DIVERSITY

Starting in 1978 with the Supreme Court decision in the *Regents of University of California v. Bakke* (438 U.S. 265 (1978)), a fundamental shift occurred in the application of affirmative action, from redressing and ameliorating racial discrimination and societal inequality to defending diversity as a compelling state interest. In the *Bakke* case, the Supreme Court deemed the use of racial quotas as unconstitutional and an example of reverse discrimination in violation of the Fourteenth Amendment and Title VI of the Civil Rights Act of 1964. In this ruling the University of California at Davis medical school's admissions policy was declared unconstitutional because it set aside 16 out of 100 slots for minority students. Although the *Bakke* ruling did indeed support the view that set-asides or quotas were illegal, Justice Lewis Powell—in an attempt to strike a compromise—nevertheless concluded that race-sensitive admissions policies were constitutional and representative of a compelling state interest if used as a "plus factor" in an effort to "obtain the educational benefits that flow from an ethnically diverse student body" (438 U.S. at 315–323).[14]

Twenty years after the *Bakke* ruling, the Supreme Court decided it was time to again make a determination regarding the use of affirmative action in higher education. In taking on both the *Gratz* and *Grutter* cases, the compromise struck by Justice Powell in the *Bakke* ruling would loom large throughout the proceedings. The Supreme Court ruling in 2003 narrowly affirmed the use of race-conscious policies by highlighting the educational benefits of a diverse learning environment. In particular, the concept of a "critical mass" of students from diverse backgrounds represented a centerpiece of the defense of affirmative action and, according to the Supreme Court, an appropriate goal for higher education institutions to pursue via their admissions policy.

In the *Grutter* case, for example, the Supreme Court heard arguments presented by the Dean of the University of Michigan Law School that claimed that without a critical mass of underrepresented minorities, students of color could "…feel isolated or like spokespersons for their race…[and potentially]…uncomfortable discussing issues freely based on their personal experiences" (288 F. 3^d at 211a). The application of "critical mass" supports the creation of a learning environment that combats marginalization of underrepresented populations. Critical mass suggests that diversity is only successful when underrepresented minority students exist within the institution in a substantial enough volume that they are able to freely engage in learning without becoming the "racial spokespersons" described by the Dean of the Law School.

This suggests a view of diversity that extends beyond minimal token representation of a population. Rather, diversity is viewed as an educational benefit shared by majority and minority students. The mission of the institution is then to represent a microcosm of the world that provides opportunities for all students to engage with others from different cultural backgrounds without feeling as though any one student represents the totality of their cultural identity. Students are able to benefit by gaining perspectives that are influenced by the variety of experiences shared by their classmates. As a result, a diverse learning environment is created where all students are able to mutually benefit from a greater community composed of cultural differences.

The relationship between diversity, leadership, and the institutional mission of higher education was also brought to the forefront by the Supreme Court's rulings in the Michigan cases. With respect to considerations regarding diversity and leadership, the Court's decision appeared to be especially influenced by several amicus briefs supplied by the U.S. military and major American businesses that stressed the "educational benefits that flow from student body diversity" (539 U.S. at 330). In regard to institutional missions, the Supreme Court's ruling in the *Grutter* case endorsed specifying how diversity is representative of a central mandate of higher education institutions beyond merely applying affirmative action in admissions.[15] By insisting that universities and colleges must provide a "highly individualized, holistic review of each applicant's file, giving serious consideration to all the ways an applicant might contribute to a diverse educational environment" (539 U.S. at 337), the Supreme Court ruling will likely compel higher education institutions in the immediate future to think more systematically about diversity as itself an educational outcome.

To this end, Patricia Gurin has surveyed Michigan undergraduates regarding their academic and social involvement with students from different ethnic backgrounds. The research revealed that significant percentages of students engaged in interracial friendships and study groups. When development over time was taken into account, students became more multicultural in their associations with others. Respondents were surveyed to determine whether many of them had at least one close friend outside of their own racial or ethnic group when they arrived at the University of Michigan. Thirty-two percent of white students and 47 percent of African American students reflected this diversity. Four years later, 46 percent of white students indicated that they had at least one close friend of color whereas 54 percent of African American students indicated having at least one close friend who was not African American.

As Gurin's report indicates, diversity makes a difference as 92 percent of white students lived in predominantly white neighborhoods prior to attending Michigan. Four years later, after experiencing diversity at the University of Michigan, white students increased their student of color friendships by 14 percent. An increase of 7 percent was also experienced among the African American student population.[16] Despite the increases in friendships across different backgrounds reported by Gurin, it is important to note that 23 percent of participating African American students reported "guarded or cautious" relations with white students and that 15 percent of participating African American students reported "tense, somewhat hostile" relations with white students.[17] Gurin, in her study of several thousand undergraduates,[18] views this tension as indicative of the need to guide students to become "conscious learners" and "critical thinkers" within a global and diverse democratic environment.[19] This, according to Gurin, can only be accomplished if the campus community supports what she terms "learning outcomes" and "democracy outcomes."[20]

Learning outcomes provide students with the means of experiencing an environment that is different from their own during a formative time of their cognitive development. Gurin refers to learning outcomes as being the product of "interaction with peers from diverse racial backgrounds, both in the classroom and informally."[21] Democratic outcomes allow students to gain a larger sense of the role they will play as a future citizen and member of a society. Gurin asserts that students who are "educated in diverse settings are more motivated and better able to participate in an increasingly heterogeneous and complex democracy."[22] Thus, both learning and democratic outcomes are dependent upon a student physically experiencing diversity. Through the formal environment of the classroom and less formal interaction with peers, students are able to challenge their perspectives and explore behavioral and psychological diversity. According to Gurin, it is precisely because of the social diversity at the University of Michigan that students are able to experience "the discrepancy, discontinuity, and disequilibrium that are so important for producing the mode of thought educators must demand from their students."[23]

But in the case of the University of Michigan, two important considerations have seriously diminished the research findings surrounding diversity and the successful defense of affirmative action. The first is the notion of a critical mass of minorities, so pivotal to the legal defense of race-conscious policies, which constituted at the time of the suit only a tiny percentage of African Americans (8 percent) and

Hispanics (5 percent) out of the total undergraduate population of the University of Michigan.[24] The second is that in a 2006 referendum the voting public of Michigan eliminated the use of affirmative action in the state.

With respect to higher education, the message seems loud and clear: when it comes to competition over valuable degrees from prestigious institutions, the benefits of diversity have not translated into significant increases in the rates of minority access to selective universities or colleges, nor have they convinced a predominantly white and older electorate of the need to continue supporting race-conscious admissions policies. Instead, the benefits of diversity appear to be contradictorily correlated with a scarcity of students of color and highly restrictive admissions criteria that especially restrict access to elite institutions for African Americans and Hispanics, who constitute a growing share of the nation's poor.

RACE AND SCARCITY VALUE IN HIGHER EDUCATION

Ironically, the scarcity of minorities at selective institutions has served as the primary rationale for the preservation of affirmative action. For example, in furnishing extensive evidence underscoring the need for affirmative action, the authors of *The Shape of the River* construct a defense of race-sensitive policies contingent on the exclusivity of the institutions included in their analysis of the College and Beyond (C & B) database. More specifically, Bowen and Bok contend that one reason affirmative action is required is because nationally black students remain underrepresented at the higher SAT levels in comparison to whites and are, therefore, at a distinct disadvantage when seeking admission to selective institutions.

To test this assertion, the authors estimate the hypothetical effects of a race-neutral standard on the admissions process at five selective schools and conclude that the overall probability of admission for black applicants would significantly decline from "...its actual value of 42 percent in 1989 to a hypothetical value of 13 percent."[25] Moreover, utilizing a color-blind admissions process that exclusively relied on standardized test scores and GPA rankings would result in the overall black probability being "...roughly half of the white probability."[26] For Bowen and Bok, these findings validate the need for race-sensitive policies to both maintain a diverse learning environment and to continue to encourage selective institutions to look past numbers and consider factors above and beyond SAT scores and GPA. Factors involving the potential promise of an applicant range from their having excelled in their studies to the possible contributions they may make to professions and society at large. Those factors influencing institutional interests are the educational benefits of ensuring a wide representation of backgrounds, experiences, and talents, as well as the overall importance of maintaining strong ties to the communities traditionally served through development and alumni relations.

Surveying the data provided by Bowen and Bok, it is evident that those black students fortunate enough to gain admission to highly selective institutions experience significant labor market benefits with respect to earnings, career advancement, and overall job satisfaction. Moreover, their study also demonstrates that the benefits of affirmative action reach other areas of social and political life, especially in terms of valuable contributions to individuals' educational and personal development. In the area of leadership, for example, the authors found that black graduates of highly selective institutions are far more likely than their white counterparts to take active community and leadership roles in society.

Amongst the 1989 matriculants,

> more than 40 percent of the Black men participated in some form of community service such as working in community centers, participating in neighborhood improvement campaigns, as well as social action associations or civil rights groups. Moreover, 12 percent of Black graduates were to be found in leadership roles, which was three times as high a percentage as was found for white male respondents.[27]

Bowen and Bok do acknowledge that "...these findings [likely] reflect a demand for leadership skills that must be understood in relation to the relatively scarce supply of black men and women with comparable records of accomplishment."[28] Nonetheless, they contend that the high overall rate of participation is an indication of a strong desire and commitment by black students to "help out," especially when considering that many of the community activities taken up were of the rank-and-file variety as opposed to high-status positions of leadership.

Yet, in drawing their conclusions surrounding the benefits of race-sensitive policies at highly selective universities and colleges, Bowen and Bok can be criticized for conflating a potential trickle-down effect of individual black achievement at elite institutions with the larger and more pressing concerns undergirding the need for affirmative action in the first place: to redress past and present-day racial discrimination through the provision of wide-ranging access to educational and economic opportunities. Focusing on access for African Americans, for example, at the five schools primarily used in the Bowen and Bok study highlights the limits of affirmative action at elite institutions. Out of a total of 40,000 applications for admission to fill 5,166 places for the 1989 cohort, a little more than 2,300 (or approximately 5.8 percent) applicants had identified themselves as black. In 1989 the acceptance rate at these five schools for black applicants was 42 percent, compared to whites, which was 25 percent, meaning that out of 2,300 or so black applicants, roughly 966 students were offered admission.[29] On the surface this appears to be a favorable outcome, especially when taking into consideration the scarcity of places available, as approximately 18.7 percent of the admitted students were black. But despite their significant presence in the admitted category, blacks only comprised 7.1 percent of the actual share of the entering '89 class. In other words, out of the 40,000 applications, less than 400 black students (approximately 1 percent of the total applicants) ended up attending the five institutions in question.

As these figures convey it is apparent that at the heart of Bowen and Bok's argument involving affirmative action lies the following contention: without race-sensitive means it would be practically impossible for elite universities and colleges to secure an acceptable black yield,[30] since the degree of selectivity characterizing these institutions is derived from hierarchical rankings based on a combination of GPAs, class rank, and SAT scores, and low acceptance rates.[31] However, what remains obfuscated in Bowen and Bok's defense of diversity and affirmative action is an underlying contradiction involving the selectivity of elite institutions: race-sensitive policies (as currently conceived) actually reinforce the exclusivity of elite institutions because the use of lofty threshold standards (SAT, class rank, GPA, and low rates of acceptance) can be maintained, as prestigious colleges and universities all over the country essentially compete for a very small number of highly qualified black candidates from year to year.[32]

The tendency for the pool of highly qualified black candidates to remain relatively small has been linked to what social scientists have called a black-white gap between high-achieving blacks and whites in terms of their overall combined scores.[33] Amongst the most qualified black students, whose scores are far above the national average on the SAT and are ranked in the upper echelons of all test-takers, there still exists a "black-white gap" based on the differences in standardized test scores by race. The test score gap has served as a focal point for conservatives to call for an end to racial preferences at elite colleges in part because they claim that academic standards are lowered at selective universities and colleges as a consequence of the use of so-called quotas to ensure a diverse student body. However, even when factoring in the differences in test scores between the most qualified black and white students, selective universities and colleges have not experienced an overall decline in their academic "standards" because the pool of successful black applicants is so small that there is little dampening effect on the combined average of scores of their incoming cohorts. Conversely, however, the institutional benefits derived from this small number of highly qualified black applicants is considerable as the mere presence of minority students attending predominantly white campuses buttresses the reputation of selective universities and colleges as proactive supporters and defenders of diversity and affirmative action.

But in the case of most of the universities and colleges included in the Bowen and Bok study, updated figures from 2001 indicate that the overwhelming majority of entrants at these institutions had SAT scores in the 1250–1450 range and were clearly ranked in the top half of their graduating class.[34] In contrast, 2000–01 national statistics for black graduates reveal that the average SAT scores for blacks were 430 (verbal) and 427 (math) for a combined score of 857.[35] Moreover, of the total number of black high school graduates in 2001 (392,000, or 15.4 percent of the total number of high school completers), only 55 percent actually enrolled in college (two- or four-year). Stated differently, black graduates constituted nationally 13.7 percent of all those high school completers who enrolled in college in October 2001, while representing 18.1 percent of all high school graduates who did not pursue higher education.[36] Furthermore, when looking at an increasingly influential factor underscoring admissions at elite institutions—subject matter testing (SAT IIs) and advanced placement courses—African American high school graduates remain seriously underrepresented in comparison to all other racial/ethnic groups. According to the College Board, in 2007, while 14 percent of high school graduates were black, they made up only 8 percent of those taking AP exams—and only 4 percent of those with passing scores.[37]

Although diversity is, and should be, a compelling interest in American society, affirmation of such a perspective should not come at the expense of abandoning equity and failing to address the continuation of tracking by race and class. In American higher education, this process of reproduction is primarily rooted in the tracking mechanisms that tend to separate different classes of people (by color, ethnicity, socioeconomic status, language, gender, region, etc.) and result in students from disadvantaged backgrounds attending less prestigious and well-resourced colleges.[38]

To date, despite the rhetorical affirmation of diversity expressed by four-year institutions, community colleges remain the only institutional tier in public higher education where minority students are either overrepresented or are roughly proportional to the size of their respective national populations.[39] Like most four-year institutions, community colleges have also embraced diversity in so far as recent work has focused attention on the establishment of a "new" globally oriented vocationalism resulting in the marketization of the two-year institution and its mission.[40] However, the courses offered at the two-year institutions tend to serve an entirely different function than those offered at the top-tier universities, which are steeped in the humanities and characterized by a commitment to fostering innovative and critical-analytical thinking: to enhance minority enrollments in terminal vocational-technical programs.[41] Thus while diverse students attend community colleges, the affirmation of their identities takes place in courses and programs that rarely lead to bachelor's degrees and are highly correlated with race and class.

To their credit, Bowen and Bok do not shy away from grappling with socioeconomic status (SES) issues when defending the use of race-sensitive admissions policies at elite institutions. They readily acknowledge that while 50 percent of all blacks nationally in the relevant age groups are from low SES families, only 14 percent of the black matriculants at the five elite institutions studied come from a low SES background. Conversely, the authors also observe that students from high SES backgrounds are overrepresented at elite institutions in comparison to their relevant national populations regardless of whether they are white or black. For example, "15 percent of black matriculants and nearly half of all white matriculants (44 percent) come from high-SES families—as compared with just 3 percent of the national black population and 11 percent of the national white population."[42]

Yet, these statistics, rather than highlight the benefits of diversity, expose the racial, ethnic, and class opportunity gap in higher education, a gap that continues to severely restrict the likelihood of African American and Hispanic students, who constitute a growing proportion of the nation's poor, from significantly (and proportionately to their respective group populations) improving their life chances. This reality brings us back full circle to the original thesis of the essay and raises a fundamental question: if diversity is indeed functioning as a commodity in higher education, who benefits most from the discourse of inclusion? To paraphrase Marx one final time, is the use-value of diversity disproportionately enjoyed by selective institutions and their predominantly white and high SES student bodies? In other words, has diversity in higher education become a kind of controlled experiment? One in which privileged students can experience the world (through multicultural texts, globally oriented courses, ethnic programming, cross-racial interactions, etc.) in all its diverse beauty based on a reality that is defined by a relative scarcity of minorities, especially those from harsh, segregated backgrounds?

Conclusion: Diversity and Changing the Odds

To be certain, diversity need not be a zero–sum game and I in no way intend to demonize, nor unduly criticize, predominantly white student bodies for embracing difference. After all, one of the many wonderful outcomes of Obama's historic electoral victory was the degree to which young people of all races, ethnicities, and classes clamored for change and a politics of hope over the divisiveness and fear-mongering of the prior regime. But if we are truly committed to diversity in higher education, then this commitment must be accompanied by evidence demonstrating that the practices and policies of universities and colleges lead to greater inclusion and opportunity, not just for the few, but for a significant number of students of color who are disproportionately from low-income backgrounds. Failing to provide greater inclusion and opportunity would privilege individual achievement over the much larger and more genuine transformation required to finally eradicate institutional racial discrimination.

That Obama succeeded against what must have been, only two years ago, astronomical odds, does not change the fact that as brilliant and charismatic a human being as he is, the current President remains the only exception to a longstanding and disgraceful rule regarding race in America. Equally important, his historical individual achievement must also be placed in an educational context. By context, I am referring to the difference between what Obama represents in relation to our aspirations for a better, more equitable society versus the elite educational credentials that contributed to his success as a graduate of Columbia University and Harvard Law School.

On the day of Obama's inauguration, media outlets featured countless interviews of DC public school students. These students, who are overwhelmingly African American and Hispanic, proudly proclaimed over and over again that because of Obama they now believe that it is possible one day to become President of the United States. What parent of color could not relate to such prideful and joyous expressions? But, as uncomfortable and possibly deflating as it may feel, it is imperative to also ask ourselves another question: what is the probability that the same DC students of color will enjoy the kind of educational opportunities afforded Obama? The answer is not set in stone, nor does it inevitably involve some bleak or overly pessimistic response. Rather, the answer to the question is contingent to a large degree on how far higher education is prepared to take seriously the benefits of diversity and to refuse to settle for a partial and commoditized form of inclusion.

I have outlined elsewhere some potential strategies that if aggressively implemented could result in higher education institutions, particularly the more selective universities and colleges across the nation, walking the talk of diversity in more meaningful and inclusive ways.[43] Suffice it to say here, that a first-order change requires higher education leaders, faculty, and administrators to acknowledge that to a considerable degree SAT scores remain proxies for socioeconomic background. Indeed, as Jesse Rothstein suggests, if admissions officials wish to avoid what might be called "affirmative action for high SES children" they should deemphasize the importance of SAT scores.[44] A second-order and related change is that highly selective institutions in particular should consider race and class simultaneously, as opposed to separately, in the admissions process. A case in point involves the growing use of low-income preference admissions policies at selective universities and colleges, which, due to the sheer demographic size of the white population in the United States, further disadvantages minorities because numerically there are far more poor white students with high test scores than poor students of color with high standardized test results. Yet, minority children are disproportionately (in relation to their total respective populations) living in poverty in the United States, attend on average the most underfunded and low-performing public schools, are among the most likely to be uninsured in terms of healthcare, and have far greater chance of suffering traumatic events in terms of violence, familial dissolution, or bereavement.[45]

Unfortunately, a continued reliance on standardized testing in higher education continues to function as a major stumbling block to implementing a truly inclusive (race and class) low-income policy at selective universities and colleges, one that is capable of equitably factoring in life experience in a systematic manner that would reward disadvantaged students who have overcome so much in their young lives to gain access to quality postsecondary opportunity. Inevitably, any attempt to reduce the overall importance of the standardized tests will evoke questions and concerns about academic merit. Nevertheless, universities and colleges committed to diversity beyond tokenism should be willing and able to define merit in new and more meaningful and inclusive ways. Moreover, although admissions at the institutions selected would need to be modified to provide a greater sensitivity to applicants' socioeconomic, racial, and cultural environments, elite universities and colleges would certainly benefit without compromising their standards.

Borrowing from the Texas Top Ten Percent Plan, factors in addition to academic grades and standardized test scores could be built into the admissions process. These factors could include the academic performance level of the applicant's school; responsibilities while attending school such as helping to raise children and employment; where in a given state the applicant is from; involvement in community and extracurricular activities; commitment to a particular field of study; the personal interview; whether the applicant lives in a rural, suburban, or urban area, or has attended a school under a court-ordered desegregation plan, or comes from a poor school district; whether the applicant is a "first generation" college student or is bilingual. To a degree, admissions officers at highly competitive institutions all over the country already take many of these factors into consideration on a departmental or faculty basis; however, in order to be truly effective at enhancing access these criteria should be formalized within the missions and uniform admissions practices of selective universities and colleges.

Like the Top Ten Percent Plan, government-sponsored educational opportunity programs that furnish access to higher education and academic support to low-income students could serve as a starting point for selective institutions to emulate in so far as they do not rely on high SAT scores for admissions and tend to target—albeit not exclusively—minority students.[46] However, these programs as currently structured remain far too tapered in terms of the small numbers of students admitted, especially to selective institutions, and continue to exclude large numbers of promising students who, while economically "disadvantaged," do not qualify because their families are slightly above the extremely low income eligibility requirements.[47] Nonetheless, one way for selective institutions to build on these programs and avoid punishing promising students, especially from low-income minority families who would not qualify under stringent financial requirements, is to factor in measures such as concentration of poverty and racial segregation by residential neighborhood, region, and individual schools, in addition to looking at average state expenditure per student by school district.

By incorporating such factors into the admissions process as well as factoring in average state expenditure per student by school district, selective universities and colleges would go a long way toward increasing the pool of high-potential students from disadvantaged backgrounds. Moreover, in adopting such measures, it would make it possible to acknowledge both rural poverty in areas such as the Appalachians without failing to pay close attention to heavily segregated urban areas in major cities such as New York, DC, Detroit, Chicago, and Los Angeles, as well as southern states that have consistently failed to desegregate districts, such as Mississippi, Louisiana, or Texas. Another advantage to this kind of inclusive assessment process is that focusing on a commonality of conditions or characteristics underscoring experience, regardless of whether an applicant is of African American, West Indian, African, mixed-race descent, Hispanic, or white background, would likely reduce concerns regarding ancestry that have often been raised in the context of who is or is not benefiting from affirmative action at highly selective universities and colleges.

In addition, there are alternative forms of assessment, such as the Posse Foundation's Dynamic Assessment Process (DAP), which do not rely on high standardized test scores for student selection and they need to be carefully mined by administrators, scholars and researchers, and policy makers. The Posse Foundation has identified and supported thousands of high-performing students of color from predominantly poor and urban environments who have graduated at over a 90 percent rate from selective universities and colleges. Given the movement afoot at selective public and private universities to employ holistic reviews of applicants, approaches such as the one developed by the Posse Foundation allow for factors not easily gleaned from academic records to be considered—thereby allowing selective institutions an opportunity to diversify their student bodies based on a variety of noncognitive variables and life experiences.

There is no shortage of effective models and programs that can be expanded, combined, and improved to help universities and colleges become more effective advocates of diversity. Such reforms would, especially in these uncertain financial times, require a significant investment and commitment on behalf of higher education in terms of the provision of increased financial aid and the development of comprehensive academic (counseling, mentoring, career services, etc.)

support. Yet increasing the class size at selective institutions need not bankrupt universities and colleges if the proportion of new students is carefully calculated in such a way as to not overwhelm institutional capacity.[48] Moreover, the benefits of principled and innovative leadership in relation to diversity and higher education would (1) enhance meaningful access opportunities for underserved populations and (2) provide a critical mass of underrepresented students that would allow for the gathering of data aimed at documenting their transition to college and the successes and failures experienced therein. If, for example, a pilot project was developed that involved, say, three different institutional types of select universities and colleges (i.e. Liberal Arts, IVYs, and private Research 1s) over a ten-year period, a rare opportunity exists to simultaneously support and research student achievement and diversity with an emphasis on underrepresented students.[49]

The main point of the alternatives presented here is to focus more attention on what constitutes a critical mass from the perspective of "disadvantaged" students and their experiences (racial segregation, economic hardship, familial responsibilities, deficient or underfunded public schools, etc.) as important considerations when identifying the educational benefits that flow from a diverse student body. Too often however, the group referenced when identifying what constitutes the educational benefits of a diverse learning environment are middle and high SES white students simply because they represent the overwhelming majority on the campuses of selective universities and colleges. In doing so, most of the research on diversity in higher education has missed out on an opportunity to learn from the extant literature on cultural capital and the impact of racial segregation in the United States and elsewhere. This body of work indicates the difficulties that students of color from disadvantaged backgrounds have when making the transition from context-specific knowledge associated with segregated life experiences to more critically analytic and abstract modes of understanding associated with dominant cultural perspectives and formats.[50]

The challenges to enhancing access in higher education are many indeed: they range from the continued erosion of affirmative action policies via ballot initiatives; the scaling back of developmental education at four-year institutions, the rising cost of tuition, and a concomitant decline over time in need-based aid; predatory student lending practices; and the intensification of student competition over college access. With what amounts to the worst economic recession since the Great Depression of the 1930s, postsecondary institutions are therefore confronted with difficult policy decisions regarding issues of access and equity. In the midst of these challenges is the reality that the demand for access to higher education amongst minority communities has never been greater and will continue to increase throughout the twenty-first century due to major shifts in demographics and immigration patterns.

Much has been made in higher education of the relationship between diversity and the future leadership structure of the country. It is interesting to note, however, that the association between leadership and diversity is usually cast in the future in an anticipation of the inevitable changing of the guard that is forthcoming as a result of the fundamental shift in demographics that has already but will continue even more so to define the twenty-first century. Yet, the problem of restricted access to quality higher education opportunities for students of color from disadvantaged backgrounds requires bold, transformative leadership in the present. A refusal on the part of universities and colleges to take meaningful action at a critical juncture in American history in which millions of Americans have lost their jobs, their homes, and control over their lives, will have disastrous consequences especially, albeit not exclusively, for young people of color who constitute a growing proportion of the nation's poor. Although selective universities and colleges have been impacted negatively by the economic collapse, they nevertheless remain better equipped and resourced to address the challenges of access. It is therefore time to go beyond a rhetorical affirming of diversity to identifying innovative and practical ways to reduce the opportunity gap in higher education. Failing to do so runs the very real risk of fetishizing diversity at elite institutions in the name of minority students who, for the most part, remain outside looking in.

NOTES

1. Karl Marx, *Capital* (Moscow: Progress Publishers, 1986), p. 43.
2. Ibid., p. 48.
3. For a more detailed theoretical analysis of the commoditization of race in American higher education, refer to Anderson, "In the Name of Diversity: Education and the Commoditization and Consumption of Race in the United States," *Urban Review New York* 37.5 (2005): 399–423.
4. William G. Bowen and Derek Curtis Bok, *The Shape of the River: Long-Term Consequences of Considering Race in College and University Admissions* (Princeton, NJ: Princeton University Press, [2000], c1998); *Compelling Interest: Examining the Evidence on Racial Dynamics in*

Colleges and Universities, ed. Mitchell J. Chang, Daria Witt, James Jones, and Kenji Hakuta (Stanford, CA: Stanford Education, 2003); Sylvia Hurtado, Jeffrey F. Milem, Alma R. Clayton-Pedersen, and Walter R. Allen, "Enhancing Campus Climates for Racial/Ethnic Diversity: Educational Policy and Practice," *The Review of Higher Education* 21.3 (Spring 1998): 279–302.; Patricia Gurin, University of Michigan, and Cooperative Institutional Research Program (U.S.), *Expert Report of Patricia Gurin: Gratz Et Al v. Bollinger, Et Al., no. 97-75321 (E.D. Mich.); Grutter, Et Al. v. Bollinger, Et Al., no. 97-75928 (E.D. Mich.)* (Ann Arbor, MI: University of Michigan, 1999); Smith and Associates, *Diversity Works: The Emerging Picture of How Students Benefit* (Washington, DC: Association of American Colleges and Universities, 1997).

5. David John Frank, "Rethinking History: Change in the University Curriculum: 1910–90," *Sociology of Education* 67.4 (1994): 231–242; "International Challenges to American Colleges and Universities: Looking Ahead," in *American Council on Education Series on Higher Education*, ed. Katherine Hanson and Joel W. Meyerson, 1995; David Palumbo-Liu and Hans Ulrich Gumbrecht, *Streams of Cultural Capital: Transnational Cultural Studies* (Stanford, CA: Stanford University Press, 1997); Kassie Freeman, *African American Culture and Heritage in Higher Education Research and Practice* (Westport, CT: Praeger, 1998); David John Frank, Suk-Ying Wong, John Meyer, and Francisco Ramirez, "What Counts as History: A Cross-National and Longitudinal Study of University Curricula," *Comparative Education Review* 44.1 (2000): 29; William R. Paulson, *Literary Culture in a World Transformed: A Future for the Humanities* (Ithaca, NY: London: Cornell University Press, 2001).

6. National Educational Longitudinal Study (NELS). Education Statistics, 2000.

7. Clifford Adelman, *The New College Course Map and Transcript Files: Changes in Course Taking and Achievement*, U.S. Department of Education (1999, 2nd edition); and Clifford Adelman, "Using the Modular System of Record-Keeping to Track Changes in Delivered Knowledge," paper presented at the 2001 Forum of the European Association for Institutional Research, Porto, Portugal (September 21, 2001).

8. American Council on Education, *Minorities in Higher Education 2008: 23rd Status Report* (2008).

9. Ami Zusman, "Issues Facing Higher Education in the Twenty-First Century," in *American Higher Education in the Twenty-First Century: Social, Political and Economic Challenges*, ed. Philip Altbach, Robert Berdahl, and Patricia Gumport (Baltimore, MD: Johns Hopkins University Press, 1999), pp. 109–150.

10. ACE, 2008.

11. ACE, 2008.

12. Thomas J. Kane, *The Price of Admission: Rethinking How Americans Pay for College* (Washington, DC: Brookings Institution Press, 1999); Arthur Levine and Jana Nidiffer, *Beating the Odds: How the Poor Get to College* (San Francisco, CA: Jossey-Bass Publishers, 1996); Zusman, "Issues Facing Higher Education in the Twenty-First Century."
13. Douglas S. Massey and Nancy A. Denton, *American Apartheid: Segregation and the Making of the Underclass* (Cambridge, MA: Harvard University Press, 1993); Melvin L. Oliver and Thomas M. Shapiro, *Black Wealth/White Wealth: A New Perspective on Racial Inequality* (New York: Routledge, 1995).
14. Eugene Y. Lowe, *Promise and Dilemma: Perspectives on Racial Diversity and Higher Education* (Princeton, NJ: Princeton University Press, 1999); John R. Howard, "Affirmative Action in Historical Perspective," in *Affirmative Action's Testament of Hope: Strategies for a New Era in Higher Education*, ed. Mildred Garcia (Albany, NY: State University of New York Press, 1997).
15. In addition, the Supreme Court's ruling in the *Grutter* case endorsed specifying diversity as a central feature of the mission of higher education institutions in maintaining, "The Court defers to the Law School's educational judgment that diversity is essential to its educational mission" (539 U.S. at 310).
16. Gurin, *Expert Report of Patricia Gurin: Gratz Et Al v. Bollinger, Et Al.*, Appendix E.
17. Ibid.
18. Gurin's research actually included data from the following three studies with undergraduates: (1) UCLA Higher Education Research Institute and Cooperative Research Program (CIRP) involving 9,316 students in nearly 200 colleges and universities, (2) Michigan Student Survey (MSS) involving 1,321 students at the University of Michigan, and (3) research conducted by Inter-group Relations, Community, and Conflict (IGRCC) (Gurin, 1999, 380).
19. Gurin, *Expert Report of Patricia Gurin: Gratz Et Al v. Bollinger, Et Al.*, p. 364.
20. Ibid., pp. 365–366.
21. Ibid., p. 365.
22. Ibid.
23. Ibid., p. 372.
24. Retrieved, June 23, 2004, http://www.umich.edu/~oapainfo/TABLES/FR_Prof.html#REB.
25. Bowen and Bok, *The Shape of the River*, p. 32.
26. Ibid.
27. Ibid., p. 162.
28. Ibid.
29. It is important to note that many of these students may actually be counted more than once as they were likely to be accepted at multiple institutions. Nevertheless, when looking at the actual share of black students admitted

to the five institutions (less than 400 in total), what is of greatest significance is how few African American students actually enroll.
30. The term yield refers to the difference between those students admitted versus those who actually enroll.
31. Bowen and Bok, *The Shape of the River*, p. 15.
32. By extremely low yield, I am referring to the fact that in Bok and Bowen's study, out of 996 black students originally accepted at 5 selective institutions, less than 400 ended up enrolling at these schools.
33. For a thorough explication and analysis of the differences in standardized test scores according to race, refer to *The Black-White Test Score Gap* (1998), edited by Christopher Jencks and Meredith Phillips.
34. *U.S. News and World Report*, September 11, 2001.
35. U.S. Department of Education, Digest of Educational Statistics, 2002; National Center for Education Statistics: Office of Educational Research and Improvement Washington D.C. (Report No. NCES 2003–060), Table 133, at p. 154.
36. See National Center for Education Statistics, Table 383, at p. 450. Needless to say, when looking at drop-out, high school completion, and college enrollment rates of Hispanics, similar arguments can be made.
37. *New York Times*, February 4, 2009.
38. Gregory M. Anderson, "In the Name of Diversity: Education and the Commoditization and Consumption of Race in the United States," *The Urban Review* 37.5 (2005); Gregory M. Anderson, Eleanor Daugherty and Darlene Corrigan, "The Search for a Critical Mass of Minority Students: Affirmative Action and Diversity at Highly Selective Universities and Colleges," *PEGS: Committee on the Political Economy of the Good Society* 14.3 (Spring 2006); Samuel Bowel and Herbert Gintis, *Schooling in Capitalist America: Educational Reform and the Contradictions of Economic Life* (New York: Basic Books, 1976); Kevin J. Dougherty, *The Contradictory College: The Conflicting Origins, Impacts, and Futures of the Community College* (Albany, NY: State University of New York Press, 1994); Jerome Karabel, "Protecting the Portals: Class and the Community College," *Social Policy* 5.1 (1974): 12–19; Jerome Karabel, "Community Colleges and Social Stratification," *Harvard Educational Review* 42 (1972): 521–562; David E. Lavin and David Hyllegard, *Changing the Odds: Open Admissions and the Life Chances of the Disadvantaged* (New Haven, CT: Yale University Press, 1996).
39. National Center for Education Statistics, 2001; U.S. Census Bureau, 2002; ACE 2008.
40. John S. Levin, "Global Culture and the Community College," *Community College Journal of Research and Practice* 26.2 (2002): 121–145; Levin, *Globalizing the Community College: Strategies for Change in the Twenty-First Century* (New York: Palgrave, 2001).
41. Robert A. Rhoads and James R. Valadez, *Democracy, Multiculturalism, and the Community College: A Critical Perspective* (New York: Garland Pub., 1996); *Community Colleges as Cultural Texts: Qualitative*

Explorations of Organizational and Student Culture, ed. Kathleen Shaw, James Valadez, and Robert Rhoads (Albany, NY: State University of New York Press, 1999).
42. Bowen and Bok, *The Shape of the River*, p. 48. These figures do not, according to Bok and Bowen, justify replacing affirmative action with a strict color-blind, class-based preference policy as the latter approach would result in a significant drop in the number of black students attending selective universities. I return to the issue of class-based preference alternatives in my concluding commentary.
43. Anderson, Daughterty, and Corrigan, "The Search for a Critical Mass of Minority Students," *PEGS: Committee on the Political Economy of the Good Society* 14.3 (Spring 2006).
44. Jesse M. Rothstein, "College Performance Predictions and the SAT," *Journal of Econometrics* 121.1 (2004): 297.
45. Douglas S. Massey, Camille Z. Charles, Garvey Lundy, and Mary J. Fischer, *The Source of the River: The Social Origins of Freshmen at America's Selective Colleges and Universities* (Princeton, NJ: Princeton University Press, 2003).
46. Sixty-six percent of HEOP students score below 1000 on the SATs and 81 percent of students enrolled in HEOP during the academic year of 2002–03 had cumulative averages above 2.0. Only 31% achieved averages of 3.0 or higher. Despite this gap in academic performance, HEOP graduates are highly successful students at the 57 affiliated institutions (including, e.g., New York University, Columbia University, and Cornell University). The program is able to offer students individualized attention within a comprehensive program offering academic, career, and financial counseling. In 2002–03, the program graduated 1,013 students who were predominantly black or Hispanic and had family incomes below $20,801.00. Of that population, 68% were immediately employed, bound to graduate school, or continued undergraduate education in a four-year institution. See Office of Higher Education, University of the State of New York, State Education Department, at www.highered.nysed.gov.
47. Selective colleges and universities must avoid falling into the trap of relying on the kinds of stringent guidelines employed by current opportunity programs that exclude large numbers of promising students, who, while economically "disadvantaged," do not qualify because their families are slightly above the extremely low income eligibility requirements. In New York State, e.g., students from a family of four in 2002–03 qualified for the Higher Education Opportunity Program (HEOP) only if their total income was below $27,000.
48. It is also important to note that elite institutions rarely utilize the full resources at their disposal in the admissions process to increase their intake of low-income students. For example, the work of The Institute of College Access and Success (TICAS) has long documented the extent to which selective universities and colleges do not take advantage of the

possibility to bring in significant higher PELL grants to entice and assist low-income students to attend and reduce their reliance on loans to fund their education.

49. Obviously generating larger incoming classes would entail significant infrastructural costs in terms of the building of new residences or dormitories, cafeterias, etc., as well as the hiring of additional faculty and staff (salaries, benefits, etc). Nevertheless, if Princeton can afford to provide full funding to predominantly higher SES students attending their institution, an investment to expand infrastructure and enhance academic and support system capacity is not in my view a far-fetched goal despite the reduction in the value of endowments of elite institutions such as Harvard.

50. Gregory M. Anderson, *Building a People's University in South Africa: Race, Compensatory Education, and the Limits of Democratic Reform* (New York: P. Lang, 2002); Michelle Lamont and Annette Lareau, "Cultural Capital: Allusions, Gaps, and Glissandos in Recent Theoretical Developments," *Sociological Theory* 6 (1988): 153–168; Annette Lareau, "Social Class Differences in Family-School Relationships: The Importance of Cultural Capital," *Sociology of Education* 60.2 (1987): 73–85; Annette Lareau, *Home Advantage: Social Class and Parental Intervention in Elementary Education* (Lanham, MD: Rowman and Littlefield Publishers, 2000); Annette Lareau and Erin McNamara Horvat, "Moments of Social Inclusion and Exclusion: Race, Class, and Cultural Capital in Family-School Relationships," *Sociology of Education* 72.1 (1999): 37.

10

COLLEGE ACCESS, GEOGRAPHY, AND DIVERSITY[1]

*Teresa A. Sullivan**

College officials throughout the United States have made repeated calls for diversity in their faculties and student bodies, often to the consternation of the general public. Few members of the public would object if the goal were stated in the negative: "Our college does not want to be homogeneous." Most people can readily grasp that a homogeneous campus works against the objective of intellectual broadening, which is a hallmark of a college education. Even in the decades in which a college education was limited to a small elite, the broadening objective was endorsed and sometimes sought through the "grand tour" or other travel. Thus, geography was associated with a broadened education, a fact to which I return in this essay.

The word "diversity," by contrast to the word "homogeneity," is often disparaged as code language for an unacceptable racial or ethnic preference. To be sure, diversity can be defined in terms of differing individual endowments of talent and ability. Students gifted in athletics, music, art, leadership, and other pursuits may be selected for a college class even if their grades and test scores may be somewhat lower than the norm, and this use of diversity is rarely challenged as a basis for college admissions. More commonly, however, the concept of diversity is used to describe the deliberate inclusion of people from

---

***Teresa A. Sullivan** has been Provost and Executive Vice President for Academic Affairs at the University of Michigan since 2006 and in August 2010 will become President of the University of Virginia. Previously she was Executive Vice Chancellor of the University of Texas System. She is a sociologist and demographer and served as Co-Principal Investigator of the Texas Higher Education Opportunity Project, a multiyear study of the effects of the Texas Top Ten Percent law on college admissions.

differing racial, ethnic, and linguistic backgrounds. Why this inclusion should be unacceptable arises from two different reactions to American history.

In the history of the United States, as well as the history of many other countries, group membership has resulted in group treatment. The list is long and familiar: the African American experience of enslavement and involuntary passage to the United States; the Mexican American experience of prejudice and discrimination; the nineteenth-century exclusion of Asians as immigrants; the employment discrimination faced by nineteenth-century immigrants from Southern and Eastern Europe. The heritage of these experiences included restricted civil and political rights, limited or inferior education, occupational restrictions, and poverty. Most visibly, the *de jure* segregation of Jim Crow separated African Americans from white society in the South, and some states also required the segregation of Mexican Americans and Asian Americans.

The first hostile reaction to inclusive diversity may arise from nostalgia for the old days, buttressed by anxiety about the continued entitlement of students from the majority group. As the competition intensifies for slots in the classes of the most prestigious institutions, the argument for "merit" over diversity may sound like a thinly disguised call for the most privileged students—often majority whites—to be admitted. Certainly the proponents of diversity are inclined to attribute such motives to their opponents.

The second hostile reaction, however, has an opposite origin. Many Americans reject prejudice and discrimination, and view as shameful the differential treatment of citizens because of their skin color or ethnic heritage. For at least some of these people, consideration of ethnic or racial heritage in college admissions represents a continuation, however benignly intended, of the mindset associated with either the complete exclusion of minority groups or of their limited inclusion through offensive quotas. The two-judge majority in the decision *Hopwood v. Texas*, which struck down affirmative action in the Fifth Circuit in 1996, used such reasoning to conclude that affirmative action in admissions violates the Fourteenth Amendment to the Constitution. Using this logic, any consideration of race or ethnicity by a state institution is not only suspect, but simply precluded. Similar reasoning and rhetoric underlie the state constitutional amendments adopted in California, Washington, and Michigan banning affirmative action.

The response of the diversity proponents has been that diversity benefits all students, and that the heritage of disparate treatment persists in educational disadvantage for some minority group members. The supporting arguments and research have generated hundreds of books and papers, opinion pieces and pamphlets, far more than I can review here. My objective here is different: I hope to show that the current conceptions of race and ethnicity derive from an interpretation of *geographic* diversity, what I call macro-geography, and that one solution to diversifying campuses may lie in the adoption of a strategy that emphasizes residence and school segregation, a kind of micro-geography.

THE GEOGRAPHY OF ORIGIN—MACRO-GEOGRAPHY

Scientists long ago concluded that a biological concept of race had little foundation, and more sophisticated studies using DNA have confirmed this conclusion. Regardless of skin color, humans share nearly all of their biological heritage with one another. Nevertheless, there are physical differences that distinguish broadly among populations because of long centuries of relative isolation. In themselves, these variations in skin and eye color, hair, and so on are unimportant, but they took on social significance when these long-isolated populations encountered one another, principally through voluntary or involuntary migration.

The physical variations, under conditions of a naïve ethnocentrism and xenophobia, marked group boundaries. "My people" could be visibly distinguished from "your people." Racial labels were thus applied to people whose origins were from different continents. And for much of the past two centuries, racial categories were thought to have such a deep biological basis as to mean that racial groups were radically different as a matter of nature, with the implication that differences so deep-seated must be both immutable and significant. Although today the biological rationalization has been debunked, the remaining historical, social, and cultural differences reinforce a sense of "differentness."

Ethnic distinctions were also associated with geography, although the geographic areas were typically smaller and closer at hand, such as regions or provinces. Ethnic distinctions prompted a similar we-versus-them response, but were often based on differences in language or dialect, religion, clothing, and so on. These distinctions are often indistinguishable to outsiders. For example, early immigrants to the United States from the Italian peninsula thought of themselves as Florentines or Sicilians or residents of other fairly small areas; it was in America that they learned to think of themselves as Italians, because that was how the American public, unable to make the finer distinctions, labeled them. Today, American visitors to the Balkans, for example, are unable to understand ethnic tensions nor even detect the ethnic differences without aid. Recent disturbances in Kenya, Turkey, and Central America exemplify the kinds of unrest associated with ethnic differences.

Just as these differences arise from geographic difference, a geographic solution is often proposed as a method to solve the tensions. Partition represents a kind of segregation, but also the possibility for the political control of a geographic area. Religious partition of India and Pakistan or of Ireland are familiar examples. Some observers propose a religious/ethnic partition of Iraq among Kurds, Sunnis, and Shiites as a means of reducing internal tensions.

My point is not that partition is either a good thing or a bad thing, but rather that the geographic solution to the ethnic problem reinforces my point that race and ethnicity are both social constructs with their origin in a real or putative geography. One fundamental dimension of the assessment about whether you and I are the same lies in our spatial relationship to each other and to others whom we consider similar. This real or putative origin is what I have termed macro-geography.

Empirically, macro-geography continues to have effects that are traceable in relations among some groups. Any analysis of census or representative survey data shows that on most indicators of social or economic data, there are persistent differences among groups. African Americans, in particular, continue to be residentially segregated and to have lower incomes than other American groups. Mexican Americans and Puerto Ricans have much lower levels of formal education than most other groups. Other consequences, especially in terms of socioeconomic status, are easy to identify.[2] A great deal of contemporary social science research has been dedicated to documenting and seeking to explain the persistent effects of race and ethnicity despite the legal guarantees of equality.

Normatively, many Americans believe that what I call macro-geography *should not have* any effect at all, because such effects might be discriminatory. Most nondiscrimination laws and policies explicitly identify race, ethnicity, and national origin as illegitimate grounds for discrimination. Religion, language, and skin color—traits that are often associated with race or ethnicity—are similarly forbidden grounds for discriminating. The most extreme view is that the government should never even ask for such identifiers in birth certificates, driver's licenses, or census returns, because just for the government to have the information might lead to additional abuse or targeting of groups in the future.

Advocates of affirmative action see persisting conditions of discrimination as justifying the consideration of race or ethnicity in admissions decisions, especially for purposes of representing a wide range of backgrounds and experiences in the class, but also as a means of interrupting the long-term effects of prejudice and discrimination. They would indeed use racial or ethnic identifiers in making decisions about college admissions, but they would use the identifiers for a benign purpose: providing access to a selective college.

By contrast, many Americans would argue that the use of macro-geographic indicators such as race or ethnic origin in college admissions is both unfair and illegitimate. The Supreme Court in its *Grutter* decision permitted the use of race or ethnicity as one factor among many considered in admissions decisions. Faced with referenda on this topic, however, a plurality of Americans voted against affirmative action, rejecting the use of race or ethnicity even as one factor among many in the admissions decision. Whether those who vote against affirmative action are really operating from pure motives of nondiscrimination is a matter of dispute; it is also alleged that they are really interested in reserving as many slots as possible at the most selective institutions for children of the white majority.

Regardless of the real or perceived motives for banning affirmative action, its use has been stymied in a number of states and referenda to ban it are being considered in several states. The continued use of macro-geographic indicators within the context of any government action remains controversial.

The Geography of Residence—Micro-Geography

If macro-geography produces visible results in the lives of individuals but is nevertheless deemed an illegitimate indicator in making decisions, then an alternative may be the use of micro-geography. Micro-geography refers to the smallest geographic divisions with which a person may be identified. In census data, these geographic areas might be the block or the census tract. In political divisions, the geographic areas might be the precinct. In public education, the geographic area might be the attendance zone.[3] The homes of most Americans may be characterized in terms of a number of overlapping, small geographic areas.

These small geographic areas may or may not correspond to a community. A small area such as a neighborhood may have a sense of itself, and may closely overlap attendance zones, precincts, or other divisions. Over time such a shared identity might be identified, both because of the density of everyday interactions among people who live with each other, and because they will come to share similar experiences in their treatment by others.[4] Famously, neighborhoods in Chicago were said to have developed a self-identity because sociologists at the University of Chicago had given names to the neighborhoods in a research project.

However, it is not necessary for the use of micro-geography that the people in a small area know each other or have a sense of shared identity. It is only necessary that large-scale forces at work in our economy and society tend to reinforce homogeneity within small geographic areas.[5] Residential segregation, zoning ordinances, rent controls, red-lining for mortgages and insurance, and many other institutional arrangements tend toward the same result: people live near others who are more or less similar to them. This fact is known to all marketers, who often use zip code as a short-hand method to identify the target audience for a product or service. The hit television series *Beverly Hills 90210* made that particular zip code well known for its association with affluence, but many other zip codes across the country carry a signal for marketers. Sometimes the signal is positive, such as the 90210 association with affluence, and other signals are negative, such as the association of inner-city neighborhoods with economic instability and crime.

Economics helps to explain the similarities of neighborhoods. Housing costs, which are determined by such things as the age of a neighborhood, the zoning, proximity to jobs, and so on, tend to ensure a kind of financial homogeneity among neighbors. That financial homogeneity in turn makes it more likely that the neighbors will have jobs that are similar in pay, and probably also similar in terms of responsibility, prestige, and complexity. Because education is an important determinant of jobs, the educational levels of neighbors are likely to be similar. And finally, overlay over all of these things the fact that racial and ethnic minorities are also more likely to be financially disadvantaged, and it is not surprising that in most American cities there are still neighborhoods that can be readily identified as white, black, or Latino.[6]

School attendance zones are typically organized by neighborhood, with the result that many American schools have a fairly homogeneous socioeconomic profile—and often, a fairly homogeneous racial profile as well. Throughout the United States, central city high schools are the most likely to enroll a predominantly minority student body, and suburban high schools are the most likely to be nearly all white.

So although Americans are legally free to reside in any neighborhood, and to send their children to the attendance zone for that neighborhood, a variety of social and economic forces coincide in ensuring that macro-geographies are replicated in micro-geographies. If the central city schools are also more run-down with more problems and fewer resources, then the micro-geography reinforces the disadvantage that might have been implied by the macro-geographic origins. The overlap is by no means perfect, and there are certainly mixed neighborhoods and mixed schools and mixed workplaces throughout the country, but the power of geography—as marketers can attest—remains considerable.

Interestingly, the use of micro-geography has not been found objectionable by the courts. The rights of local areas to control their own schools, even if that leads to the segregation of central city schools, were upheld in the *Milliken v. Bradley* decision. The majority opinion in the *Hopwood* decision, while striking down the use of race or ethnic origin, explicitly endorsed geographic representation: "A university may properly favor one applicant over another because of his ability to play the cello, make a downfield tackle, or understand chaos theory. An admissions process may also consider an applicant's *home state* or relationship to school alumni" (emphasis added; *Hopwood v. Texas* [1996]). It is a small step to the smaller geographic areas I have termed micro-geography.

The *failure* to consider micro-geography may actually add to the disadvantage experienced by some students. A telling example of this disadvantage is the recalculation of grade-point averages of applicants by the University of California System. Grades for honors and advanced placement courses were weighted, so that students who had taken more challenging curricula could be rewarded for their ambition and experience. Minority group advocates protested, noting that many of the central city schools attended by their children offered no such courses, so that their grades were, in effect, capped as a result of their residence and attendance zones.[7] On the other hand, it is worth considering whether the *inclusion* of micro-geography could improve access by groups that were traditionally disadvantaged in macro-geographic terms.

Improving Access by Considering Geography

Just as geography once broadened an education through travel, today geographical diversity can broaden the experience of everyone on the college campus. There are a number of ways that one might consider micro-geography in forming a freshman class. Here I discuss two prominent examples: the Texas Top Ten Percent law, and the University of Michigan's use of a marketing tool, Descriptor PLUS.

Affirmative action in college admissions was banned in Texas in 1996 as a result of *Hopwood v. Texas*. The Texas attorney general ruled that the case implied that financial aid targeted to women or minority groups was also illegal. The Texas state legislature in 1997 passed a statute now popularly known as the Top Ten Percent law, which mandated that public institutions in the state must automatically admit any student who graduated in the top 10 percent of an accredited high school in the state, provided that the student enrolled within two years of graduation and did not first enroll in any other college or university. The student decides to which schools to apply. Even though affirmative action within the confines of the *Grutter* decision is now legal in Texas, the legislature has only modified the Top Ten Percent law rather than repealing it.

The Top Ten Percent law creates an entitlement for each school district—indeed, for each high school—within the state. To the extent that attendance zones are residentially segregated, relatively more African American or Mexican American high school graduates may find themselves automatically admissible. And the large Rio Grande valley, which because of migration patterns from Mexico is predominantly Mexican American, is guaranteed that the top decile of its high school graduates can attend the Texas public institutions of its choice.[8]

The University of Texas at Austin (UT) designed its Longhorn Opportunity Scholarships to reinforce these micro-geographic aspects of the Top Ten Percent law. The scholarships are available only to top 10 percent graduates of particular high schools. These high schools are selected for specific characteristics: (1) a low fraction of the school's SAT or ACT scores were reported to the University of Texas, indicating that the high school was underrepresented within the UT student body; (2) the census tract in which the high school is located is low income according to census data; (3) the school meets a minimum size threshold. Students from the Longhorn Opportunity Schools are also eligible for all other types of merit and need-based aid, but the Longhorn Opportunity Scholarships set aside for their schools are an inducement to apply and raise the likelihood that a student will be able to afford tuition once accepted. Texas A & M University at College Station has also developed a scholarship program that is similar.

The Top Ten Percent law remains controversial because of the perceived disadvantage to private high schools and to high-quality suburban public high schools, but the racial and ethnic diversity of the UT student body has returned to approximately the same level as under traditional affirmative action in the pre-*Hopwood* days. In addition, an unanticipated consequence was the addition of rural white students to the student body, most of them from high schools that had not previously been represented in the freshman class. While not a perfect solution, the targeting of admissions and financial aid to smaller geographic areas—in this case, high schools and their surrounding census tracts—was at least partially successful in improving access to the state's flagship public universities.

The efforts in Michigan are more recent and their success remains to be seen. The Michigan electorate amended the state constitution in November 2006 to ban the use of race, ethnicity, and gender in college and university admissions. The wording of the ballot initiative closely tracked the wording of Proposition 209, which was adopted in California in 1995. The University of Michigan is seeking to use information about whether applicants' high schools or residential neighborhoods are underrepresented in the student body. This information is applied in a holistic admissions evaluation that examines many aspects of a student's academic experience, extracurricular activities, and other accomplishments.

The tool being used, Descriptor PLUS, is marketed by the College Board and uses clustering algorithms of the type now common in many marketing applications. The student's high school is characterized in terms of the composition of its student body, and the student's neighborhood (determined from residential address) is characterized in terms of the composition of the neighborhood as revealed through census data.[9]

A wealth of information about these smaller geographic areas is potentially available for such algorithms. For the school, for example, there is information about the fraction of students eligible for reduced-price lunches, the rate of success on accountability examinations, the resources per student, and so on. Depending on the state, the agency that oversees secondary schools may also provide information about high school graduation rates, the fraction of the school in college preparatory studies, the fraction of the senior class taking calculus, the proportion of students who are bilingual, and many other things that might be relevant indicators of disadvantage. The College Board has the advantage of its proprietary database from the millions of students from every high school who have registered for its exams.

Data on residential neighborhoods are typically produced from U.S. Census Bureau sources. The decennial census provides information on geographic areas down to the city block. These data are carefully aggregated and the reports edited so that no information can be used to identify a household. Even with the editing constraints, however, the census data permit identification of a wide range of characteristics of both housing and households, such as average housing value, general condition of housing, household incomes, size of households, proportion of immigrants, average age, and so on. From other sources, a variety of other indicators could be constructed. For example, disease rates and health indicators vary by small geographic area.[10]

Such indicators are not simple substitutes for race or ethnicity. The majority of the population is also likely to be the majority of most subgroups in the country. Thus, most of the people in poverty are white, even though the likelihood of being in poverty is greater for African Americans or for Latinos. But because of the geographic clustering of the population along socioeconomic lines, there is also some correlation with race and ethnicity.

These indicators are also not foolproof guides to socioeconomic status. The poorest neighborhood may have the eccentric neighbor with extensive assets not hinted at by a family's modest home. An affluent immigrant family may prefer for reasons of language and culture to remain in a neighborhood with others whose salaries are much lower than theirs. Especially in neighborhoods undergoing rapid change, families may represent a wide range of backgrounds and any assumption of homogeneity will fail. Nevertheless, the use of small geographic groups—what I have called micro-geography—as an additional indicator in admissions may provide some additional source of diversity when conventional affirmative action is not permitted.

Conclusion

The positive claim that diversity is needed on campus encounters resistance, but campus leaders will persist in their efforts to diversify the campus because of the benefits diversity provides all students. Employers are coming to realize that a diverse work group can be more creative in problem-solving, and a similar analysis applies to college classrooms.[11] Legal restrictions on tactics such as affirmative action do not diminish the need for diversity but they do make diversity harder to achieve by the most efficient means, which is the direct consideration of race, ethnicity, or other characteristics.[12]

The consideration of micro-geographic origins, such as the composition of neighborhoods and high schools, offers one proxy means for increasing the diversity of public universities. Even increasing the number of high schools represented within a freshman class represents an important means of strengthening a university's links to its publics, and may contribute at least some of the diversity that is currently sought through affirmative action.

NOTES

1. This research was supported by grants from the Ford, Mellon, Hewlett, and Spencer Foundations, and NSF (GRANT # SES-0350990), with institutional support from the Office of Population Research, Princeton University (NICHD Grant # R24 H0047879). Some of the data used for this study are restricted and, therefore, not available from the authors.
2. Douglas S. Massey, *Categorically Unequal: The American Stratification System* (New York: Russell Sage Foundation, 2007).
3. Allan C. Ornstein, *Class Counts: Education, Inequality, and the Shrinking Middle Class* (Lanham, MD: Rowman & Littlefield Pub. Group, Inc., 2007).
4. William J. Wilson and Richard P. Taub, *There Goes the Neighborhood: Racial, Ethnic, and Class Tensions in Four Chicago Neighborhoods and Their Meaning for America* (New York: Knopf, 2006).
5. Howard Frumkin, Lawrence Frank, and Richard Joseph Jackson, *Urban Sprawl and Public Health: Designing, Planning, and Building for Healthy Communities* (Washington, DC: Island Press, 2004).
6. John Iceland and Rima Wilkes, "Does Socioeconomic Status Matter? Race, Class, and Residential Segregation," *Social Problems* 53.2 (2006): 248.
7. Sunny Xinchun Niu, Teresa Sullivan, and Marta Tienda, "Minority Talent Loss and the Texas Top 10 Percent Law," *Social Science Quarterly* 89.4 (2008): 831.
8. Kim Lloyd, Kevin Leicht, and Teresa A. Sullivan, "Minority College Aspirations, Expectations, and Applications under the Texas Top 10% Law," *Social Forces* (March 2008): 1105–1138.
9. "Descriptor PLUS," *College Board* (2007), http://professionals.collegeboard.com/higher-ed/recruitment/descriptor-plus.
10. Frumkin, *Urban Sprawl and Public Health*, ed. Lawrence D. Jackson and Frank Richard.
11. Scott E. Page, *The Difference: How the Power of Diversity Creates Better Groups, Firms, Schools, and Societies* (Princeton, NJ: Princeton University Press, 2007).
12. Douglas Laycock, "The Broader Case for Affirmative Action: Desegregation, Academic Excellence, and Future Leadership," *Tulane Law Review* 78.6 (June 2004): 1767–1842.

11

HIGHER EDUCATION AND THE CHALLENGE OF INCLUSION

Marvin Krislov[*]

With the election of the first president of color, the question of the need for affirmative action to assist racial and ethnic minorities once again has come into the forefront. The irony of Senator Obama defeating Senator Hillary Clinton (a white woman) for the Democratic nomination has been noted by many. Although affirmative action programs in the United States are more commonly seen as focusing on people of color, programs both public and private remain that are targeted toward professions or studies where women are underrepresented (such as math, science, and medicine). The electoral success of now President Obama and former Senator Clinton could be viewed as undercutting the argument for affirmative action. Indeed, some of the discussion concerning Mr. Obama posits that Americans may be driving toward a world where race is irrelevant. Such arguments are undercut by data in the 2008 election demonstrating that race and ethnic identity continue to have meaning, and that discrimination has far from disappeared.[1]

In other ways, the electoral success of President Obama and Senator Clinton underscores the importance of ensuring that people of color and minorities actually attend and graduate from this nation's

---

[*] **Marvin Krislov** became the fourteenth President of Oberlin College in 2007, where he is also a professor in the Politics Department. Mr. Krislov served as Vice President and General Counsel at the University of Michigan from 1998 to 2007, where he was also an adjunct professor for the Law School and the Political Science Department. He led the University of Michigan's legal defense of its admissions policies, resulting in the 2003 Supreme Court decision recognizing the importance of student body diversity.

most elite institutions. Senator Obama attended Occidental College, Columbia University, and Harvard Law School; Senator Clinton attended Wellesley College and Yale Law School—all five are private, highly selective institutions where affirmative action programs have been and continue to be supported.[2] One might argue that success in American society, particularly for women and people of color, is related to the ability to attend that class of institutions with the most affluent and powerful networks.[3] Moreover, one could argue that the financial support necessary to mount a credible national campaign derives from the candidates' ability to tap into such networks.

At the same time as this seemingly bright development, there are cautionary reports. While the number of racial and ethnic minorities in the United States is increasing,[4] significant income and wealth gaps remain between racial and ethnic groups, as well as continued residential segregation.[5] Although women—and in particular, white women—have made enormous advances in populating American higher education, pockets remain where women are underrepresented—notably medicine, math, and the sciences.[6] Thus, affirmative action advocates might point to this underrepresentation (as well as historical discrimination) as justifying higher education admissions, aid, and support programs that target racial and ethnic minorities, and, in some instances, women.

On the other hand, those who dislike such programs invoke the ideology of "merit" and propose substituting class-based programs. Other programs, such as the Texas Top Ten Percent program, may have the ability to improve the representation of racial and ethnic minorities in certain selective institutions, but for many institutions affirmative action continues to be critical to creating an inclusionary campus. This particularly applies to campuses with national student populations where guaranteed admission by high school would not work.[7] Given the leaky pipeline in so many of our nation's high schools, it may be necessary to focus on underrepresented groups as a way of ensuring sufficient numbers of students of color. (Additionally, there may well be concerns about increasing faculty and staff of color, as well as women, in certain underrepresented fields.) In this essay, I discuss the legal and political framework for colleges and universities who choose to employ a variety of affirmative action policies, and then consider future policy directions and responses.

The Supreme Court's 2003 decisions in the Michigan cases created challenges and opportunities for colleges and universities seeking to enroll diverse student bodies. A majority on the Court recognized the importance of student body diversity (and racial inclusion) in colleges and universities and this recognition seems likely to survive the changing composition of the Court. (Justice O'Connor, who wrote the seminal majority opinion in *Grutter*, has since left the Court, but the Court's recent decisions in the *Parents Involved in Community Schools v. Seattle School District No. 1* and *Meredith v. Jefferson County Board of Education* suggest that the Michigan decisions will remain intact.) The 2003 cases came after a decade of political and legal wrangling over affirmative action policies,[8] including adverse legal rulings in *Adarand v. Pena* and *Hopwood v. Texas*, and California and Washington State initiatives barring any such state sponsored programs. The Michigan cases created a legal framework, consistent with the Supreme Court's 1978 *Bakke* decision, for colleges and universities to consider race and ethnicity in student body admissions. The Michigan cases did not deal with the consideration of gender in admissions nor did they delve into other higher education programs such as financial, outreach, or support programs. The cases did not touch on employment or contracting. Perhaps most significantly, because the lawsuits turned on the role of race and ethnicity, the ensuing popular discussion gave relatively short shrift to other considerations in admissions, including, for instance, socioeconomic disadvantage, athletic prowess, alumni connection, or geographic diversity.

What do the cases mean for colleges and universities? First, they establish a legal framework that applies to admissions decisions in any public institution (under the equal protection clause of the Constitution) and to any private institution that receives federal funds (under Title VI, the legal standard is equivalent to that of the equal protection clause). Relying on previous jurisprudence, the Court held that strict scrutiny review applies to any consideration of race or ethnicity in admissions. Any plan that includes race must meet two separate elements: compelling governmental interest and a "narrowly tailored" structure. In the Michigan cases, the Court upheld the concept that student body diversity qualified as a compelling educational interest. The Court relied on testimony from educational experts and, perhaps most importantly, from outside validators—including business and military leaders. These leaders argued that experience in multicultural settings better prepared students for leadership and competition, decreased discrimination, and promoted success in global settings. The compelling interest thus derived from an educational imperative rather than from the notion of redress or remedy for past discrimination. Court jurisprudence has limited the reach of remedial programs, although clearly historic and continuing discrimination against persons of color is relevant to the context of racial and ethnic residential separation in which so many Americans still live. As Justice O'Connor's majority opinion stated, "...race unfortunately still matters."[9]

The *Grutter/Gratz* cases, taken together, also outline what constitutes acceptable tailoring, or construction of admissions plans. Thus, the University of Michigan's undergraduate admissions process did not pass muster, because the point system (assigning points for qualities including academics, test scores, extracurriculars, and socioeconomic status and underrepresented minority status) was too "mechanistic" or "formulaic" in its consideration of race and ethnicity. The law school's admissions process commanded majority support because it was "holistic," not assigning any particular weight to racial or ethnic identity, and considered underrepresented minority status on a clearly individualized basis. The law school process thus resembled the Harvard undergraduate process described by Justice Powell in his *Bakke* opinion.[10] Specifically, the court criteria for narrow tailoring[11] require that a program (1) not use a quota or "insulate a category from competition,"[12] (2) look for reasonable, race-neutral alternatives,[13] (3) not unduly harm members of any racial group,"[14] and (4) terminate the program after some amount of time.[15] The Supreme Court was satisfied with the law school's method—it took race as one of many factors that spoke to an applicant's background, offering that applicant a perspective useful to other students. The law school did aim to achieve a "critical mass," but, as the Court found, not to satisfy any one number, but to ensure that there were enough students of various backgrounds to encourage real conversation among students.[16]

The Michigan cases, therefore, provide a framework for creating legally defensible admissions policies at selective schools. We know that most American colleges and universities are not selective, although the most selective campuses tend disproportionately to produce leaders in many professions.[17] In part because the Michigan cases and much of the public debate focused on racial and ethnic considerations in admissions, there may well be a false sense that racial and ethnic identity are largely determinative of admissions at competitive schools. In the 2008 presidential election, for example, Senator McCain suggested that racial quotas were still an issue in American higher education.[18] The 1978 Supreme Court decision in *Bakke* explicitly outlawed racial quotas or set asides in college and university admissions. In the Michigan cases, there was no evidence at all that quotas or set asides were employed. Instead, the Court affirmed the validity of the concept of "critical mass" being used to suggest that the educational benefits required more than a token, or isolated, number of minority students.

Second, while the evidence before the Supreme Court attempted to contextualize the relative weight of factors including racial and ethnic identity, that picture has probably not been properly conveyed to many in the public. For those selective institutions, other considerations have far greater weight. In addition to academic qualifications (such as grades and, where applicable, standardized test scores), many colleges and universities place weight on athletic prowess[19] and alumni status. Indeed, as works such as Dan Golden's *The Price of Admission*[20] and Peter Schmidt's *Color and Money*[21] depict, for many colleges and universities, nonacademic factors that could hardly be labeled "meritocratic" play into admissions decisions in ways that may tend to disadvantage students from less privileged backgrounds.

In response to the concern that less affluent students of all races and ethnicities have less opportunity to attend selective colleges and universities, Bill Bowen and others have recently advocated for greater attention to including students from less affluent backgrounds.[22] As many have pointed out, increased emphasis on socioeconomic factors (sometimes called economic affirmative action) will not substitute for race-conscious admissions in ensuring racially and ethnically diverse student bodies.[23] Nevertheless, greater focus on socioeconomic background may well serve important societal goals, particularly for public institutions.[24] After Prop 2 passed (in effect barring the consideration of race and gender in public institutions), the University of Michigan revised its undergraduate application to allow an applicant greater opportunity to provide information, on a voluntary basis, about his or her family's socioeconomic background. The recent movement of more affluent colleges and universities to eliminate loans as a form of aid for all, or for a portion, of their student bodies may also broaden the family backgrounds of incoming students.[25] Many public institutions have established significant scholarship programs for needy students. Ultimately, greater focus on support for underprivileged populations may well increase minority and first-generation enrollments. However, the recent economic downturn may jeopardize some need-based aid.[26] Part of the challenge for American higher education lies in the leaky pipeline, particularly in many public schools, and in the less than unified approach to preparing students for a competitive college admissions process.[27]

The cases then help guide admissions options in selective colleges and universities under the federal Constitution. Yet state laws also apply to public institutions. Fueled by former California Regent Ward Connerly, California's Prop 209, Washington State's I-200, and Michigan's Prop 2 have sharply curtailed those states' public universities' ability to consider race and ethnicity in admissions decisions. In 2008, these initiatives had only minimal success—several did not make it onto the ballot. Colorado became the first state to reject the affirmative action ban, while Nebraska did pass it.[28] In response to such political initiatives, some competitive universities have developed what have been called "race neutral" methods of promoting diversity in response to legal or political constraints. Following the ban by the UC Regents and then the ensuing enactment of Proposition 209, California developed a method of ensuring placement of the top 4 percent of high school graduates in a UC campus. This has had mixed results, according to observers including the former President of the UC System.[29] Just recently, the continued sharp decline in African American enrollment at UCLA and at its law school has provoked much controversy.[30]

For many colleges and universities, even more important than affirmative action in admissions are financial aid efforts. Some programs may be broadly based, not targeted toward students of color explicitly, although students of color may benefit significantly because they may tend to come from less privileged families. However, many colleges and universities have found targeted financial aid a valuable tool in recruiting and retaining students of color. California officials interpreted Prop 209 to prohibit race-conscious aid. To address this challenge, alumni recently established a separate foundation to target scholarships toward African Americans.[31] Although the Michigan cases dealt with admissions, not financial aid, some believe that the cases suggest that scholarships where race or ethnic identity is one factor among many can be acceptable. Others believe that the Michigan cases cannot support gender- or race-exclusive scholarships or aid (such as, e.g., a scholarship targeted to only Latino students or only women). There are few court cases dealing with this topic, although the 1994 Fourth Circuit decision in *Podberesky v. Kirwan* underscores the need for scrutiny of public aid programs open only to those of certain racial or ethnic backgrounds.[32] Nevertheless, the 1994 Department of Education guidance has not yet been superseded or revised and it suggests that such programs (including in some instances race- or gender-exclusive cases) may in fact be defended as promoting diversity.[33] Other colleges and universities have adapted aid programs in keeping with one "holistic approach," as endorsed by *Grutter*.

At the same time, many colleges and universities have intensified efforts to provide financial aid to students of lower-income backgrounds. These programs, including the North Carolina Covenant, the Michigan M-Pact program, and the Oberlin Access Initiative may vary in details, but it is clear that the rising cost of college (and the perception of its unaffordability) has created demand for greater economic access. The Longhorn Scholarships in Texas, promoted as part of the percentage plan solution, have opened the Austin flagship to a broader group of students.[34] These financial aid programs, particularly if well publicized and well supported, can bolster public support for inclusionary programs. Additionally, programs such as Posse and Questbridge, that recruit highly qualified students in urban public high schools for scholarships to competitive colleges and universities, help advance inclusionary goals by increasing both racial and economic diversity. Posse, which has been enormously successful at schools including Oberlin, features a leadership program that enhances the educational experience of its participants.[35] For the vast majority of American colleges and universities, limited resources will constrain financial aid choices for the foreseeable future. Thus, the trade-offs between targeted or nontargeted aid, merit aid vs. need-based aid, and in-state vs. out-of-state residents, will likely continue.

Additionally, many colleges and universities engage in outreach programs to recruit and retain students. Special fly-in weekends and other programs for prospective and/or high school students often target high-achieving students of color and, sometimes, women, in underrepresented fields such as math or science. While these programs are likely to withstand legal scrutiny, the Hi-Voltage decision[36] by the California Supreme Court suggests that in California at least, post-Prop 209 recruitment efforts targeted toward students of color may be questioned.

Thus, the legal environment constrains admissions processes at highly selective institutions, but the Michigan cases provide a road map for those institutions that wish to pursue race or ethnic diversity through a holistic process. The logic of those cases may apply to gender-conscious admissions, although those discussions have received far less attention. One might expect that targeted financial aid programs might draw more attention, with outreach programs seemingly the least likely to attract controversy. Indeed, the political environment (not only with regard to ballot initiatives but with regard to state legislatures, boards of trustees, and the public) is likely to be challenging with regard to many affirmative action programs, particularly admissions.

Much of the preceding discussion has focused on creating a diverse student body—through admissions, targeted aid, and outreach. Yet another topic—more administrative and policy oriented—is how to promote effective inclusionary programs on campus. Critics such as Justice Scalia in his dissent in *Grutter* decry what some see as largely segregated campuses.[37] For many reasons, this argument rests on largely anecdotal evidence that may be exaggerated. Beverly Tatum discusses the natural and unsurprising circumstance in which the significance of groups of visible minorities may seem more prevalent on some majority-white campuses.[38] But the need for supporting group identity may surface in many contexts—religion, ethnicity, gender, athletic ability, politics, to name just a few—without undercutting the sense of a broader community.

On many campuses, and certainly on Oberlin's campus, faculty, staff, and students discuss and debate academic and cocurricular policies that try to balance the needs of constituency groups (particularly minority groups) and the desire to create a unified community. As American campuses become more religiously diverse, questions of observance of religious holidays will be ever more salient as will academic offerings on world religions. In the context of race and ethnicity, one might expect that an increasingly diverse American population will lead to increased focus on serving underrepresented groups. At Oberlin College, many programs support diversity and inclusionary activities, including the Kosher-Halal Co-Op that brings together Jewish and Muslim students and their friends. The Office of Residential Life at Oberlin sponsors many activities inspiring interaction among different students. The Multi-Cultural Resource Center and the Office of Spiritual and Religious Life aim to promote cross-cultural and interfaith dialogue. The Oberlin College Dialogue Center (OCDC), under the Office of the Ombudsperson, trains students to facilitate constructive dialogue, especially where there is a potential for conflict. During my tenure so far as President, the OCDC has facilitated discussions on topics including police-student-town relations (and allegations of racial profiling), and the role of the military. In the curricular realm, Oberlin requires students to satisfy a "cultural diversity" requirement. Additionally, Oberlin has allowed limited credits for approved student- and nonfaculty-taught courses in the Experimental College, which broadens curricular options.

International students also offer particularly rich opportunities for cross-cultural exchange. At Oberlin, for example, during my first year, international students organized academic panels on Pakistani and Iranian elections and the political situation in Burma. International students at Oberlin present an annual program entitled "Myths and Truths about My Country," which allows students from a range of countries to offer personal perspectives that enrich the community's understanding.

These examples, to touch on only a few, demonstrate how a diverse campus can build bridges and institutional structures. Some campuses have created offices or vice presidents of diversity; others have chosen to ask administrators to prioritize this area. There is no simple formula for stimulating the optimal learning environment of diverse students, faculty, and staff populations, but it seems clear that such efforts will require ongoing attention at the highest level of higher education administration. Not only pipeline programs, but campus outreach and inclusion must also be a priority.

As we look to the future, we must recognize that, even as/though we have elected our first president of color and have made enormous strides in many ways, we continue to face largely segregated and uneven public schools, an increasingly diverse domestic population, and limitations (legal, financial, and others) in creating higher educational institutions that are truly inclusive.

What policies might higher education, and its supporters in government and the private sector, pursue to address the pipeline problem and the growing population of young people of color in this country? First, efforts to make college affordable can certainly help, although these efforts have undoubtedly been hurt by the economic downturn and the rising costs of higher education.[39] We might expect the federal government to consider policies aimed at increasing aid, perhaps through a combination of direct funding and tax credits.

One of the challenges to "fixing" the access problem in the United States is the enormous range of problems, across states, and across public and private entities. This not only creates structural challenges, but also communications challenges. Higher education may not appear to have a unified or coherent message regarding its own inclusionary programs. Given the mix of private and public institutions, the multiplicity of messages is to some extent inevitable. However, the higher education community can and should promote better public understanding of, and support for, inclusionary policies and programs.

To promote this goal, colleges and universities should examine admissions, aid, and outreach policies to determine whether they are in fact achieving their goals. Many of us see student bodies that are by and large affluent and not as diverse as might be desired for educational reasons. Colleges and universities should also work to become more transparent about costs and programs. With the overlapping federal, state, and institution-specific aid programs, it is no wonder that most families are confused about the true cost of college. The process of selective admissions decisions, understandably, baffles many families. In this dialogue, it is also important to address many lingering myths about affirmative action and inclusionary programs in higher education. One might wish to talk about the importance of composing a class, pool constraints, and competing priorities. It may be crucial to emphasize the role of such programs in providing access to women and the socioeconomically disadvantaged. Public education campaigns, particularly if they can be coordinated among institutions, may address popular misconceptions.

This work can be focused on campuses and systems. Ultimately, though, success may depend on collaboration across colleges and universities, with community colleges and a broad array of K-12 and philanthropic, corporate, and governmental institutions to address gaps. The Michigan cases helped stimulate a broad coalition of support for the importance of diversity to American—and global—society. These partnerships should be nurtured and developed. In particular, one might anticipate that higher education will work more closely with the K-12 sector to improve college preparation.

As we look forward, we can anticipate that inclusionary programs, including those labeled as affirmative action, may take on a great symbolic significance in the legal, economic, and/or political arenas. Higher education must defend and explain these programs, their context, and their interrelationships, if we are to maintain the confidence and trust from our various constituencies. To do so will require an understanding that higher education is a public good, not simply a benefit to individuals. The election of President Obama may help break down barriers, but on addressing the education achievement gap and on creating truly inclusionary campuses, there remains much more to be done.

NOTES

1. P. Healy, "Beneath Campaign Surface, Obama's Race Remains a Potent Issue," *The International Herald Tribune* October 13, 2008; but see Kate Zernike and D. Sussman, "For Pollsters, the Racial Effect that Wasn't," *The New York Times*, November 6, 2008, www.nytimes.com/2008/11/06/us/politics.
2. All five institutions signed amici briefs supporting the position of the University of Michigan in the U.S. Supreme Court cases of *Grutter v. Bollinger* and *Gratz v. Bollinger*. See Brief for Harvard University et al. as Amici Curiae Supporting Respondents, *Grutter v. Bollinger*,

539 U.S. 306 (2003) (No. 02-241); Brief for Harvard University et al. as Amici Curiae Supporting Respondents, *Gratz v. Bollinger*, 539 U.S. 244 (2003) (No. 02-516); Brief for Columbia University et al. as Amici Curiae Supporting Respondents, *Grutter v. Bollinger*, 539 U.S. 306 (2003) (No. 02-241); Brief for Columbia University et al. as Amici Curiae Supporting Respondents, *Gratz v. Bollinger*, 539 U.S. 244 (2003) (No. 02-516); Brief for Amherst College et al. as Amici Curiae Supporting Respondents, *Grutter v. Bollinger*, 539 U.S. 306 (2003) (No. 02-241); Brief for Amherst College et al. as Amici Curiae Supporting Respondents, *Gratz v. Bollinger*, 539 U.S. 244 (2003) (No. 02-516). Needless to say, Wellesley College stands out in this group in its commitment to the education of women since that has been and continues to be its mission. *Wellesley College*, "The College: An Introduction," http://www.wellesley.edu/Welcome/college.html (last visited September 13, 2008).
3. On President-Elect Obama's advisor's links to elite institutions, see, e.g., D. Brooks, "The Insider's Crusade," www.nytimes.com (November 21, 2008).
4. S. Roberts, "A Generation Away, Minorities May Become the Majority in U.S.," *The New York Times* August 14, 2008.
5. See U.S. Census Bureau, *Housing and Household Economic Statistics Division, Racial and Ethnic Residential Segregation in the United States: 1980–2000*, August 11, 2008. See also G. Witte and N. Henderson, "Wealth Gap Widens for Blacks, Hispanics," *The Washington Post*, October 18, 2004.
6. D. Castelvecchi, "Numbers Don't Add Up for U.S. Girls," *Science News* 174.10 (November 2008): 10. See also N. Andrews, "Climbing through Medicine's Glass Ceiling," *The New England Journal of Medicine* 357.19 (November 2007): 1887–1889.
7. Marta Tienda, Sunny Niu, and Teresa Sullivan, "Minority Talent Loss and the Texas Top 10 Percent Law," *Social Science Quarterly* 89.4 (December 2008): 831–845.
8. Affirmative action may be defined as programs encouraging and/or providing special consideration for women and/or minorities in such areas as contracting, hiring, and education. This article focuses primarily on affirmative action or inclusionary programs in higher education.
9. *Grutter v. Bollinger*, 539 U.S. 306, 21 (2003).
10. *Regents of the University of California v. Bakke*, 438 U.S. 265 (1978).
11. Usefully summarized in *Comfort v. Lynn School Comm.*, 418 F. 3d 1, 17 (3rd Cir. 2005).
12. *Grutter*, pp. 315–316.
13. Ibid., p. 339.
14 Ibid., p. 341.
15. Ibid., p. 342.
16. As the Court noted, the number of minority offers of admission varied considerably from year to year. Ibid., p. 336.

17. See, e.g., Justice O'Connor's discussion of the legal profession in *Grutter*.
18. D. Jackson, "McCain Sides with Ban on Affirmative Action," *USA Today*, July 27, 2008.
19. See, e.g., Bill Bowen's work showing that athletic consideration exceeds the weight given to racial and ethnic identity. Sarah A. Levin and James L. Shulman, *Reclaiming the Game: College Sports and Educational Values* (Princeton, NJ: Princeton University Press, 2003). See also "Race-Sensitive Admissions: Back to Basics," William G. Bowen, The Andrew W. Mellon Foundation Annual Report (2002), http://www.mellon.org/news_publications/annual-reports-essays/presidents-essays/race-sensitive-admissions-back-to-basics (last visited October 9, 2008).
20. Daniel Golden, *The Price of Admission: How America's Ruling Class Buys Its Way into Elite Colleges—and Who Gets Left Outside the Gates* (New York: Crown Publishing, 2006).
21. Peter Schmidt, *Color and Money: How Rich White Kids Are Winning the War over College Affirmative Action* (New York: Palgrave Macmillan, 2007).
22. William G. Bowen, Martin A. Kurzweil, and Eugene M. Tobin, *Equity and Excellence in American Higher Education* (Charlottesville, VA: University of Virginia Press, 2005).
23. In fact, although families of minority students are disproportionately represented in the bottom income quartile, their research suggests that if "class-based" admissions preferences were completely substituted for race-sensitive admissions preferences, minority student enrollment would be expected to fall substantially (in most cases reduced by about half, regardless of whether the institution is public or private). Ibid., pp. 183–186. This would occur because white people still constitute the vast majority of underprivileged people. Commentators such as Bowen et al. advocate that attention be paid to race and ethnicity, as well as to socioeconomic status.
24. On the value of diversity to institutions and organizations, generally, see Scott E. Page, *The Difference: How the Power of Diversity Creates Better Groups, Firms, Schools, and Societies* (Princeton, NJ: Princeton University Press, 2007).
25. Oberlin, e.g., eliminated loans for those students whose families are eligible for federal Pell Grants. Harvard and some of the better-endowed colleges and universities eliminated loans for all students.
26. Geraldine Fabrikant, "Colleges Struggle to Preserve Financial Aid as Investments Decline," *The New York Times*, November 10, 2008.
27. "America's Promise Alliance Launches National Campaign to Combat Nation's High School Dropout and College-Readiness Crisis," April 1, 2008, Press Release, www.americaspromise.org; See also Editorial Projects in Education Research Center, *Cities in Crisis: A Special Analytic Report on High School Graduation*, April 1, 2008, Released via the America's Promise Alliance, http://www.americaspromise.

org/uploadedFiles/AmericasPromiseAlliance/Dropout_Crisis/SWANSONCitiesInCrisis040108.pdf.
28. R. Wiedeman, "Analysis: How Colorado Became the First State to Reject a Ban on Affirmative Action," *The Chronicle of Higher Education*, November 10, 2008.
29. Richard Atkinson, "Diversity: Not There Yet," Op-Ed, *Washington Post*, April 20, 2003. See also "Undergraduate Access to the University of California after the Elimination of Race-Conscious Policies," Report by Student Academic Services, Office of the President, March 2003, available at http://www.ucop.edu/sas/publish/aa_final2.pdf.
30. "Professor Protests over Black Admissions at U.C.L.A.," *The New York Times*, August 30, 2008. See also David Leonhardt, "The New Affirmative Action," *The New York Times*, September 30, 2007.
31. R. Trounson, "Scholarship Fund to Help Blacks Go to UCLA," *Los Angeles Times*, March 29, 2007.
32. *Podberesky v. Kirwan*, 38 F.3d 147 (4th Cir. 1994).
33. Appendix 5: U.S. Department of Education Final Policy Guidance, *Nondiscrimination in Federally Assisted Programs; Title VI of the Civil Rights Act of 1964*, Final Notice, Federal Register, Vol. 59, No. 36, February 23, 1994. (In particular, Principle 4: "Financial Aid to Create Diversity"), available at http://www.ed.gov/about/offices/list/ocr/docs/racefa.html (accessed February 5, 2010); see also U.S. Department of Education, "The Use of Race in Postsecondary Student Admissions," August 28, 2008, www.ed.gov/about/offices/ocr/letters/raceadmissionspse.html (accessed February 5, 2010).
34. M. Long and M. Tienda, "Winners and Losers: Changes in Texas University Admissions Post-*Hopwood*," *Education Evaluation Policy Analysis* 30.3 (September 2008): 255–280.
35. See Rassan Salandy, ed., *The Posse Foundation 2007 Annual Report*, The Posse Foundation (Worchester: Saltus Press, 2007).
36. *Hi-Voltage Wire Works, Inc. v. City of San Jose*, 24 Cal. 4th 537 (2000).
37. *Grutter v. Bollinger*, 539 U.S. 306 (2003) (Scalia, J., dissenting).
38. Beverly Daniel Tatum, *"Why Are All the Black Kids Sitting Together in the Cafeteria?": A Psychologist Explains the Development of Racial Identity* (New York: Basic Books, 1997).
39. Fabrikant, "Colleges Struggle to Preserve Financial Aid," found at www.nytimes.om/200811/11/giving; T. Lewin, "College May Become Unaffordable for Most Americans, Report Says," New York Times, December 3, 2008, found at www.nytimes.com/2008/12/03/education.

12

NOTES FROM THE BACK OF THE ACADEMIC BUS

*William A. Darity, Jr.**

Racial/ethnic groups who are far more absent than present on college and university faculties face a variety of complex circumstances while entering careers as professors. I find it useful to construct a typology of departments based upon their record of inclusion and exclusion of black scholars. Borrowing from the notion of "sundown towns"—towns where no blacks were allowed to be present after sunset—I label faculties that have never had a black colleague as sundown departments. Faculties that have had at least one black faculty member in the past but have none at present can be called midnight departments. Finally, those faculties with a single black colleague are window dressing departments. It is straightforward to apply the typology to economics departments, the field among the social sciences that has proven to be the most resistant to altering their demography to include black scholars.

Sundown departments in economics include the University of West Virginia, the University of Chicago, the University of Minnesota, the University of California at Santa Barbara, the California Institute of

---

*William A. ("Sandy") Darity, Jr. is Arts and Sciences Professor of Public Policy Studies and Professor of African and African American Studies and Economics at Duke University. He previously served as Director of the Institute of African American Research, the Moore Undergraduate Research Apprenticeship Program, the Undergraduate Honors Program in economics, and Graduate Studies at the University of North Carolina. He is a past president of the National Economic Association and the Southern Economic Association. He has also taught at Grinnell College, the University of Maryland at College Park, the University of Texas at Austin, Simmons College, and Claremont-McKenna College.

Technology, the University of California at San Diego, the University of Florida, Emory University, the University of Georgia, Johns Hopkins University, Mississippi State University, the University of North Carolina at Greensboro, Clemson University, the University of South Carolina, Virginia Polytechnic Institute, George Washington University, and Wayne State University. Midnight departments include the University of Texas at Austin, Stanford University, the Massachusetts Institute of Technology, Boston University, the University of California at Riverside, and Notre Dame University. Window dressing departments include the University of North Carolina at Chapel Hill, Duke University, Georgetown University, Brown University, Harvard University, and Princeton University. Yale University with *three* black faculty members is an absolute rarity among departments at institutions that historically have had a predominantly white student body.

The worst of these departments are those that not only have no black faculty but also have a shabby record of producing black PhDs. At least some of the departments listed earlier—MIT, Stanford, UC Berkeley, UNC at Chapel Hill, and Clemson—have a recent track record of increasing the supply of black economists. But many others maintain academic segregation not only by their failure to hire black colleagues but also by their failure to expand the pool of black economists through their own graduate programs. In general, economics departments have "diversified" toward greater inclusion of scholars from other countries and toward greater inclusion of female scholars. But inclusion of black scholars, especially those who are U.S. citizens, has proven to be particularly halting in economics. Indeed, Gregory Price has demonstrated that, paradoxically, as the supply of new black doctorates in economics has gone up the probability of a PhD-granting economics department hiring a black economist has gone down.[1] This would suggest that the problem of racial exclusion in economics is not merely a pipeline problem, that is, the absence of sufficient numbers of black scholars completing PhD programs, but also a discriminatory resistance to their employment as faculty colleagues.

An additional symptom of the racial sentiment in the field of economics is the stunning evidence concerning the racial composition of the editorial boards of the seven journals published by the major organization of economists, the American Economic Association. The seven journals are the *American Economic Review, Journal of Economic Literature, the Journal of Economic Perspectives,* and four new American Economic Journals: *Applied Economics, Economic*

Policy, Macroeconomics, and *Microeconomics.* Associated with these seven journals are approximately one hundred and seventy positions as editors, associate editors, coeditors, and editorial board members. *None* of the economists holding these positions is black. Regardless of whether this was an unconscious demographic oversight or a deliberate action, it is at least as unconscionable as the fact that the first female recipient of the Nobel Prize in economics received the award 50 years after it was given initially.

Conditions at the American Economic Association journals contrast with the composition of editorial boards for the leading journals in other social science disciplines. The *American Sociological Review* (ASR), the journal of the American Sociological Association, lists France Winddance Twine among its editors, and Prudence Carter, John Sibley Butler, Orlando Patterson, and Cedric Herring among its editorial board members. Indeed there is a sufficient number of black sociologists having an editorial role at *ASR* to yield a group embodying substantial ideological diversity—from Butler and Patterson on the right to Twine and Carter on the left. Admittedly, though, the editorial board for the *American Journal of Sociology* based at the University of Chicago strongly resembles the demographics of the boards for the American Economic Association journals.

The American Political Science Association's journal, the *American Political Science Review,* includes Claudine Gay as an editor and Robert Gooding-Williams and Melissa Nobles on the editorial board. The journal of the American Anthropological Association, *American Anthropologist,* has about fifty members on the editorial board, two of whom are Irma McClaurin and Karla Slocum, both black anthropologists. Only economics has achieved the distinction of a null set of black economists on the editorial boards of the journals published by its major organization.

* * *

The coin of the realm of an academic career is publications. The only insulation scholars have, particularly if they are engaged in ideologically challenging research or if they are from a group that is presumed to be cognitively inferior, is to build a strong publication record. In the humanities and in some branches of the social sciences—especially anthropology and political science—books published by academic presses are valued. But in sociology, and especially in economics, articles in refereed journals are what matters. Moreover, there is a ranking of journals that leads article placement to take an important role in how a scholar's portfolio is evaluated. Some departments actually provide their faculty members with a list of periodicals in their field and indicate how many points literally will be scored by getting a hit in a particular journal.

In economics the typical list will locate journals such as the *American Economic Review*, the *Economic Journal*, *Econometrica*, the *Quarterly Journal of Economics* (QJE), the *Journal of Political Economy* (JPE), and the *Review of Economics and Statistics* at the top. These are so-called "general" journals that ostensibly consider papers in all areas of economics in contrast with the so-called "specialty" journals that accept work in narrower subfields of economics, for example, the *Journal of Human Resources* for labor economics or the *Journal of Economic History* for economic history. A few of the "specialty" journals—such as the two just mentioned—are held in very high regard on the usual ranking list, but the most prestigious journals are, more often than not, "general" journals.

The QJE and the JPE also function as "house" journals. They are intimately connected to the economics departments at Harvard and at the University of Chicago, respectively. For a scholar to place a paper in either one of these journals it is vital to have a connection to one of those departments, either as a faculty member, as a former graduate student, or as a presenter at a seminar there. So these two journals while highly prestigious are not genuinely "open" journals.

The openness of any journal is contingent on the perspective and practices of the editor, associate editors, and editorial board. A cautious editor committed solely to "normal science" will be reluctant to take seriously submissions that do not conform to conventional standards of quality and acceptability. More adventurous editors are more likely to be found heading journals that are not among the very top ranked. They will take greater risks with the papers they accept, sensing that perhaps one of those more creative papers somehow will have an impact and affect the direction of inquiry in the field. It only takes one or two papers from a less prestigious periodical to make a splash over the course of three to five years to alter the profile of the journal.

What is tricky here is the fact that the articles in the more prestigious journals tend to have a wider audience. Many more economists will at least scan the cover of the *American Economic Review* to determine whether there are one or two articles they actually will read in a given issue than say, one of my favorite journals, the *Journal of Socioeconomics*. This is partially because the *American Economic Review* automatically goes to all members of the American Economic

Association as a benefit of their dues payment. But it is also because the stature of the journal means that the workaday economist can convince himself or herself that they have a sense of what is happening in the field by at least reviewing the *Review*. They may never know that far more interesting and provocative articles are being published in the *Journal of Socioeconomics*.

To the extent that the assessment of a scholar's portfolio is not solely dictated by the placement of their articles but also by citation count, the authority of the existing array of highly ranked journals is reinforced. Since more people see them, placing an article there raises the odds that the article will have a higher than average citation count. The brave editor of the less prestigious journal is hoping that the madly original article he or she has decided to accept and publish will make waves and garner citations, but it will have to do so without a boost from the journal itself. The author of the madly original article probably tried initially to place it with Big Journal but failed and, following the wise rule that it is better to publish somewhere than not to publish at all, turned to Small Journal. With some self-promotion and serendipity the article in Small Journal may still gain wide visibility. But it is less likely.

For black scholars oriented toward a mainstream research agenda, placing papers in the top journals largely will be a question of producing work that is technically of a high quality and of developing the networks that will give their work a real hearing at the higher ranked journals. For the black economist who is challenging orthodoxy things are more difficult. Indeed, they become especially difficult if the black economist is not only challenging orthodoxy but doing research on race.[2]

Jewish scholars studying their people or Asian scholars studying their people do not tend to run into the same skepticism and resistance for their work as black scholars studying their people. If the black scholar produces a paper that says black-white disparities are due primarily to internal black cultural or behavioral dysfunctionality the article may get published readily, even in a premier journal. If the black scholar produces a paper that concludes that black-white disparities are primarily attributable to white racism and the operation of white privilege, he or she probably will have to move from Big Journal to Small Journal to get a fair hearing for their work. Indeed, I would contend that a white scholar reaching similar conclusions would have better odds of placing the work in Big Journal.

Two anecdotes are pertinent here. First, I recall a referee report that Rodgers and Spriggs received when they had a paper rejected at

Big Journal that subsequently was published in the *Review of Black Political Economy*.[3] The referee said pointedly that a major problem with their paper was the fact that they seemed to be assuming that blacks and whites had equal intelligence. Second, I recall submitting a paper to Big Journal that reported on estimates of the magnitude of racial discrimination in American labor markets between 1880 and 2000 and having the paper returned to me by the editor without it even being sent to referees, allegedly because it was not of sufficient "general" interest for the readership of the journal.

A raw careerist might have reacted to these types of rebuffs by altering what their research says. I do not profess to have high integrity in all arenas, but my research program is one where I try to maintain my allegiance to the pursuit of truth as strongly as I can. So an opportunistic change in position is not an option for me. But I can continue to attempt to place work of mine that is less charged—work that is not necessarily on racial inequality—in Big Journal while finding a home for my more charged research with more receptive editors at less highly ranked journals. Of course, such a strategy requires double duty—doing work on both fronts simultaneously. The black scholar's additional burden?

* * *

Black scholars who do obtain faculty positions must be willing to move. At minimum they must be willing to move until they receive tenure; they may still need to be willing to move thereafter. This advice applies to nonblack scholars as well, but there is a uniquely precarious position faced by black scholars. Things happen to black scholars that do not happen to nonblack scholars. I know firsthand of two instances during the past academic years where black assistant professors undergoing second-year, pretenure reviews were told that the reviews are pro forma, everyone gets a positive one, and there was nothing to worry about. In both cases the young scholars received brutal reports from their department chairs, detailing expectations of them that went far beyond the accomplishments of the existing faculty. In one case the faculty styles themselves as politically progressive and certainly would be stunned by any intimation that their actions are racist. My advice to both of the junior scholars is to move.

After all, the other item of currency in the world of the university is mobility. It is valuable for your department to know that there are other places that want you. Of course, a scholar's degree of mobility is closely related to their research portfolio. Publications or imminent publications are vital to this process.

I think that every assistant professor should be on the market the year before their tenure decision is made by their department and university, for at least two reasons. First, the odds of receiving an external offer are greater if one is in the market before the tenure decision is made. A negative tenure decision creates a scarring effect that may make it much more difficult to obtain an offer from another department. Second, having an external offer in hand increases the likelihood that a scholar will receive tenure at their home institution.

Keep in mind, though, that once you enter the market you have to be willing to leave. If your home institution does not at least match the offer you received from another school, it is time to go.

* * *

The exclusionary practices that are so pronounced in the field of economics have an obvious subtext—the implicit (and occasionally explicit) belief in the inferiority of the black scholar. I warrant that the rhetoric that would be used to justify the complete absence of black scholars from the editorial boards of the American Economic Association journals would have something to do with the notion that the economists invited to serve were all selected on the basis of "merit." This is simply another instance of the fetishization of "merit" as a rationalization for discriminatory outcomes. It presumes that there would be no improvement in the quality of the editorial practices of the journals and ultimately their content if the composition of their editorial boards were different. Other disciplines in the social sciences have begun to reach a different conclusion. Economics remains the most backward discipline on this score.

In this context, it is worth noting how many key theoretical concepts would be missing from the social sciences, especially economics, if not for the contributions of black scholars. Just to name a few: discounted dynamic programming, stereotype threat, oppositionality, surplus labor, the dual economy, the plantation economy, color-blind racism, programmed retardation, modernity, blaming the victim, educational subnormality, deficit models, resiliency, social capital, legacy effects, neocolonialism, postcolonial melancholia, double consciousness, preemptive extermination, racialization, and cultural representation.

NOTES

1. Gregory N. Price, "The Problem of the 21st Century: Economics Faculty and the Color Line," *Journal of Socioeconomics* 38.2 (2009): 331–343.
2. Patrick I. Mason, Samuel L. Myers, Jr., and William Darity, Jr., "Is There Racism in Economic Research?" *European Journal of Political Economy* 21 (2005): 755–761.
3. W. M. Rodgers and W. E. Spriggs, "What Does the AFQT Really Measure?: Race, Wages, Schooling and the AFQT Score," *Review of Black Political Economy* 24.4 (June 1996): 13–46.

13

CONSTRUCTING JUNIOR FACULTY OF COLOR AS STRUGGLERS: THE IMPLICATIONS FOR TENURE AND PROMOTION[1]

*Stephanie A. Fryberg**

Shortly after I became an Assistant Professor in the Psychology Department at the University of Arizona, my graduate advisor emailed me a speech by the President of Princeton University, Shirley M. Tilghman. The speech, "Changing the Demographics: Recruiting, Retaining, and Advancing Women Scientists in Academia," focused on the underrepresentation of women in science and engineering. President Tilghman, a microbiologist, espoused the benefits of increasing diversity for the sciences and for the country more generally, all the while emphasizing the academy's "moral obligation" to change:

> For every girl who dreams of becoming a scientist or engineer, there is a moral obligation on our part to do everything we can to even the playing field so her chances rest on her (dare I say innate?) abilities and her determination, just as it does for her male counterparts. It is not sufficient to shrug our shoulders, invoke all the historical reasons for the situation, call upon the leaky pipeline, or bemoan the difficulty of changing culture.[2]

*** Stephanie A. Fryberg** is an Assistant Professor in the Department of Psychology and an Affiliate Faculty member in American Indian Studies at the University of Arizona. Her primary research interests focus on how social representations of race, culture, and social class influence the development of self. In 2007, Dr. Fryberg was the recipient of the Society for the Psychological Study of Social Issues (SPSSI) Louise Kidder Early Career Award for contributions of research to society.

Whether the goal is to attract the best and brightest young minds, to increase the range of problems being studied, or even to fulfill a moral obligation to even the playing field, she argued that the academy must be "eternally vigilant" to the societal images that work against increasing the diversity of the academy.

When I was invited to write this chapter, I immediately thought about President Tilghman's speech. I was reminded of one comment in particular: that when she closes her eyes and thinks about a "stellar scientist" or "the best person for the position" she can imagine a woman, whereas her male colleagues in the sciences or other male administrators close their eyes and often see only a man. Her comments profoundly, yet simply, illuminate the tacit, and sometimes not so tacit, cultural ideas and stereotypes (e.g. the image of the stereotypical scientist) that shape the perceptions, communicated expectations, and decisions of administrators and faculty in the academy.

Just as President Tilghman's speech highlights the role that cultural stereotypes play in the professional careers of women in the sciences, this essay focuses on how the culture of the academy, in particular the cultural stereotypes of junior faculty of color, influence the personal and professional lives of junior faculty of color. The essay is composed of three parts. First, I describe how the culture of the academy influences what university administrators and senior faculty envision as the contributions of junior faculty of color. Second, using a short personal vignette and then drawing on social science research, I unpack the *struggler phenomenon*, the widely held but largely unacknowledged belief that junior faculty of color will struggle in the academy. Third, I offer strategies for junior faculty of color and for university administrators and senior faculty to reduce the effects of the struggler phenomenon.

Images of "Successful Academics"

Images of success in the academy depend, in large part, on the culture of the academy. This culture of the academy consists of the implicit and explicit patterns of ideas, values, and practices that emerge over time (i.e. they are historical products), but that are widely shared and tacitly instantiated in the everyday functioning of the academy.[3] The culture of the academy gives meaning and structure to everyday activities and sets up guidelines for rewarding different ways of thinking, feeling, and acting. For example, the culture provides guidelines for evaluating "good" scholarship, teaching, and service, and as such it provides specific ideas about the value and meaning of the choices and the behaviors of "successful" academics.

In other words, to be successful, like any other behavior or activity, requires partaking in culturally specific meanings and practices.[4] If the culture of the academy is set up to foster particular understandings of what it means to be successful, then the ideas and behaviors of people participating in the academy will reflect these ways of being.[5] For instance, when university administrators and senior faculty think about "successful academics" at most large research universities, they think about individuals who are highly productive (i.e. publish or perish), independent thinkers (i.e. separate from graduate advisors) who generate novel and highly reputable scholarship. Faculty who resemble this characterization are most likely to be deemed "successful."

These traditional ideas about what constitutes successful academics are also fostered by university policies and practices used to evaluate faculty.[6] The guidelines for promotion and tenure, for example, may vary depending upon the size and teaching focus of the university, but they are nonetheless derived from and reinforced by culturally specific sets of ideas about success. Senior faculty, for instance, serve as cultural gatekeepers for the university by sitting on promotion and tenure committees. Given their own past success with promotion and tenure, they are led to believe that their own experiences reflect the successful academic ideal. Consequently, they are likely to use their own paths as models for how future junior faculty should conduct themselves.

These cultural ideas and stereotypes about faculty success are reinforced by the racial-ethnic composition of the academy. For example, given the relative homogeneity of senior faculty (i.e. white, male), there is a tendency to uphold traditional standards of excellence as though these standards are impartial evaluative tools, rather than products of the academy's historical practices of exclusion.[7] The divergence between the images of successful academics and the images of faculty deemed as "falling short" (often junior faculty of color, women) is, in effect, fostered by these traditional standards of excellence. The culture shapes what people see (i.e. what is good) and, importantly, what people do not see (i.e. what is different) in evaluating junior faculty. If images of successful academics *do not include* images of junior faculty of color, then just as President Tilghman's colleagues could not see women as "stellar scientists," administrators and faculty alike will not recognize the contributions of junior faculty of color to the academy or to society and, as a result, will not see them as tenure-able. Instead, the contributions of junior faculty of color are measured by how much the individual approximates the "mainstream" culture of the academy rather than by how much they generate new theories and ideas, create innovative pedagogy, and model pathways of success for students.

Moreover, in order to meet the ongoing efforts to diversify the academy, administrators and faculty must begin to recognize how the cultural ideas and stereotypes, such as images of "successful academics," match the perspectives of and thus privilege the dominant group, while also potentially undermining those who are historically underrepresented in the academy. The problem may not be that junior faculty of color are falling short or that they are not excellent, but rather that universities, as they currently operate (i.e. with a definition of success that matches the dominant group), are unable to see junior faculty of color as excellent.

THE STRUGGLER PHENOMENON

When junior faculty of color do come to mind, they are constructed as strugglers, as individuals who are likely to strive, but likely to fall short of the high standards of the academy. This predicament, where junior faculty of color are seen as potential strugglers, but not as potentially successful academics, is what I refer to as the *struggler phenomenon*. Specifically, in the context of the academy, the struggler phenomenon refers to the widely held, but largely unacknowledged belief that junior faculty of color will struggle in the academy. I call this belief a phenomenon because it is a pervasive, but not an entirely concrete, experience that requires explication. While some junior faculty of color will inevitably struggle because they are ill prepared (they may have received poor mentoring, for instance), the struggler phenomenon is about the construction of junior faculty of color as strugglers in the absence of objective evaluation. The struggler image elicits concern for all constituencies—junior faculty of color, university administrators, and senior faculty—that junior faculty of color might struggle, but the burden of the concern is most likely to fall on junior faculty of color. To overcome the image, junior faculty of color must spend time and energy vigilantly trying to disprove the possibility.

By illuminating the struggler phenomenon, I highlight the ways in which the cultural ideas and stereotypes about junior faculty of color create tacit barriers on the path to promotion and tenure that ultimately undermine a university's efforts to increase the representation of junior faculty of color in the academy and to make explicit the contributions of junior faculty of color to the university and to society more generally. In order to unpack the struggler phenomenon and to highlight both the experiential and the theoretical power of the struggler phenomenon, I first provide a brief personal narrative about the struggler phenomenon and then I use social science research to further unpack it.

Personal Narrative: The Struggler Phenomenon

I am currently an Assistant Professor in the Department of Psychology and an Affiliate Faculty member in the American Indian Studies Program at the University of Arizona. Prior to these positions, I completed my doctorate at Stanford University in 2002 and then worked at Stanford as an Assistant Dean for Multicultural Graduate Student Services in the School of Humanities and Sciences for nearly two years. I am a social psychologist, but my affiliations with the Center for Comparative Studies in Race and Ethnicity (CCSRE) at Stanford University and the Future Minority Studies (FMS) Research Project, a multi-institutional consortium of scholars interested in minority identity, education, and social transformation, gave me a strong interdisciplinary scholarly focus.

In addition to my academic identity, I am an American Indian woman. I grew up on the Tulalip Indian Reservation in Washington State, where I developed a strong sense of collective responsibility and a very real and at times painful understanding of social inequality. I am the first person on both sides of my family to go to college and I am the first and only person from my tribe to receive a PhD. Throughout most of my academic career, I felt like an imposter in a world that was made for other people. Nonetheless, I worked hard to develop the academic skills and the cultural competence needed to successfully navigate between my life as an academic and my life as a member of my tribe. I persisted, in large part, because of my deep commitment to social justice and social change, particularly with respect to Native issues, and because of supportive mentors who helped me envision a future for myself in the academy.

Now, as an Assistant Professor, my life largely centers on the University of Arizona (UA). As I turn to these experiences, I want to offer a short caveat about the UA. I do not believe that the university is unique in its struggles with diversity-related issues, nor do I believe that my experiences as an Assistant Professor at the UA are unique (i.e. especially bad or especially good). In fact, while my experience has at times been difficult, my general impression is that the university, in particular my department, demonstrates great willingness to learn and to change. Moreover, from my untenured colleagues all the way up to the university president, I have found genuine support for diversity initiatives, including those initiated by me. This is generally more than I can say for my junior faculty of color colleagues at other institutions.

When I first arrived at the UA, I felt well prepared for the professoriate. What surprised me most was how quickly this feeling changed. I cannot pinpoint any one incident that caused this change, but rather, I believe that it was the accumulation of many smaller insidious events and experiences. For example, in my first year, the sheer number of warnings, some subtle and some not, I received from university administrators and senior faculty about junior faculty of color struggling to get tenure at the university made me acutely aware that I was, in their eyes, a potential struggler. On one particular occasion, a senior colleague and I met with a high-level administrator in our college; we wanted to learn more about diversity efforts on campus. At the very outset, I was explicitly told that I could not get involved in diversity efforts because "'women and minorities do not do well here."

This incident is not an isolated experience, nor is it the worst experience I have had as a junior faculty member. Yet, this experience is a clear example of the complexity of the *struggler phenomenon*. On the one hand, the experience surprised my senior colleague and me. We both took many days to digest the experience, but more importantly, it took many months to fully understand the interaction. In hindsight, I believe the administrator had good intentions. The belief that junior faculty of color engage in too much service and not enough scholarship is pervasive in the academy. The problem with the comment, however, was twofold: first, the administrator saw me through the lens of stereotypes about women and junior faculty of color, and, second, the administrator heightened my concern about being seen as a struggler—as one of the women or minorities who will not do well at the university.

The perception of junior faculty of color as strugglers, as not producing enough research and, thus, as struggling to attain tenure, shaped my junior faculty experience from the very beginning. Initially, I wanted to compensate for my potential struggler identity by publishing quickly and often. I tried to supplement my research with quicker and faster projects. My primary research focuses on improving the academic and psychological well-being of American Indian children and young adults, but conducting this research in Native communities is a very slow process and recruiting a sizable Native college sample is an even slower one. My attempt to augment my research with other populations unfortunately proved to be equally difficult. I had little motivation or passion for this research and trying to split my time between projects ultimately slowed the whole process, further exacerbating my growing fear of being seen as a struggler.

The next stage of the publishing process also contributed to this fear. In the field of social psychology, American Indians are rarely studied. Consequently, there is little direct theoretical or empirical research on which to build my arguments or to support my data. Nonetheless, given the outstanding contributions of my field on issues related to stereotyping and prejudice and to culture and self-understanding, I was confident that the field was right for me and that it had much to offer Native issues. Unfortunately, I was not prepared for the insensitivity and/or ignorance on the part of some reviewers. In one paper, for example, we argued that American Indian students hold different models of education (i.e. they have different ideas about the purpose and function of education) than Asian American and European American students.[8] One of the reviewers offered as an alternative explanation that "perhaps American Indians just have less innate mental ability than Asian Americans or European Americans." Another reviewer, on yet another paper, asked that we replicate the research with another minority group because it is "not clear what we can learn from studying American Indians." These examples proved more the standard than the exception, and more importantly, they formed the criterion by which the manuscripts were rejected.

While critical reviews are not uncommon in the field, I found that these types of reviews hurt me personally and served as reminders that people like me (i.e. American Indians) do not belong in the field. The reviews and rejected manuscripts also came to symbolize an additional invisible burden; I would have to educate my field about American Indians and I would have to do so as a junior faculty member under the evaluative pressure of tenure. Each time I received one of these reviews, I coped by putting the review aside for weeks or months until I found the mental and emotional energy to contend with it. In the end, my manuscripts were not getting published; my sense of belonging in the field and in my university was decreasing along with my passion and motivation for conducting research; and my fear of being seen as a struggler was growing.

By the beginning of my third year, the psychological toll surpassed the professional toll. Initially, I was concerned about being seen as a struggler, but by my second year *I began to see myself* as a struggler.

This change disoriented me. I interpreted normal research setbacks, such as rejected papers or negative reviews, as evidence that I did not belong. Annual evaluations and mid-tenure review evoked tremendous anxiety, and asking for help with manuscripts and grant writing evoked negative physiological responses (e.g. stomach pains, sleepless nights). By the end of my third year, even asking people I trusted evoked these responses. Eventually, I stopped caring what my colleagues thought. I dismissed review letters and any evaluative comments, positive or negative.

In the end, I engaged in a popular strategy for protecting one's self. I coped with the situation by becoming less identified with my field, department, and university.[9] Disidentifying helped me to survive; it helped me to reconnect with my preprofessorial self and to see the ways in which my growing fear of being seen as a struggler was changing me. This recognition made the struggler phenomenon salient and illuminated the factors that helped me to confront these feelings of struggle. Before moving on to the social science research, I want to highlight three of these factors—mentors, professional affiliations, and disidentification.

Mentors

While the role of mentors in the successful promotion of junior faculty is widely acknowledged,[10] I offer here specific information about the roles and the advice my mentors provided. The two mentors who played the most crucial roles in my junior faculty experience were my graduate mentor and my senior faculty mentor.

My graduate mentor, Dr. Hazel Rose Markus (Professor of Psychology; Director of the Center for Comparative Studies on Race and Ethnicity), helped me develop the academic and professional skills necessary to be a professor, but she also helped me to transition into being an assistant professor. Moving away from my family and tribal community for school was difficult, but the assumption was that the move was temporary. Taking a job further from home seemed more permanent. While my commitment to the community remained, I felt like, at least symbolically, I was turning my back on my community. Without the ongoing connection with my graduate advisor, I might have left the academy, but she provided a sense of stability and connection and offered valuable advice about how I could address the needs of my community through scholarship, by being a professor.

In this sense, Dr. Markus encouraged me to follow my own path—to achieve the goals that were most important to me. Of course, she noted at various times that a more traditional path in the professoriate would be safer, but she never lost sight of the fact that my commitment to my tribe and to social justice for Native people nationwide were central to my passion and motivation for research, teaching, and service. Rather than question these commitments or ask me to wait until I achieved tenure, which is perhaps the worst piece of advice university administrators and senior faculty give to junior faculty of color, she encouraged me to combine my research interests with my tribal commitments, and she created the space to envision such a path.

While I cannot speak for all junior faculty of color, the reason I believe asking junior faculty of color to wait to help their communities until post-tenure is terrible advice is that it implies a lack of understanding for what it means to be a person of color from an underrepresented and underserved community, and because it inadvertently functions to remind junior faculty of color why the academy is not for people who are motivated first and foremost by social change. For example, each year in my tribal community a number of people die from drug and alcohol abuse, depression, suicide, and various other health discrepancies. Moreover, the majority of children drop out of school, which further places them at risk for these problems. If I am the only person from my community with a PhD, how can I turn my back on them for six years? People are dying! From an early age, my elders encouraged me to get an education and to come back and help our children. I endure the daily struggles as an academic in hopes that I might one day live up to their wishes.

One might argue that the academy is not the right place for someone with my motivations, but the issues in my community are not just local issues—they are national issues. Most universities have as their mission to serve the people of the state. I truly believe that the academy—to the extent that it is the academy of all people and all communities in this country—needs more people with these collective motivations in order to be socially relevant and to meet the needs of these diverse communities. The academy, assuming I survive the tenure process, is a place where I can, in theory, do both. Asking me, however, to put my personal success ahead of my community's needs undermines my motivation for the academy. When I receive such advice, I feel an instant need to distance myself from the person offering it and am left to wonder, once again, whether the academy is the right place for me. Dr. Markus, in contrast, valued the diversity of experience I brought to the academy and convinced me that, while it would be difficult, I could fulfill the expectations of both the academy and my tribal community.

The second mentor was my senior faculty mentor. The UA requires junior faculty to select a senior faculty mentor early in their first year. I chose Dr. LouAnn Gerken (Professor of Psychology and Linguistics; Director of Cognitive Sciences) because I participated in a workshop she led on grant writing for junior faculty in the College of Social and Behavioral Sciences. Since my interaction with senior faculty was, at that time, fairly limited, I selected Dr. Gerken because I thought her style of interacting with participants engaging and her style of discussing the politics of grant writing ingenious. I could imagine talking to her about sensitive issues and I anticipated that she would be open to my perspective. I was soon rewarded for trusting my instincts.

Early in my tenure, Dr. Gerken played a vital role in my decision to stay at the UA. During my first year, a few predicaments arose which alluded to differential treatment based on race and gender. She convinced me that as my senior mentor she was the best person to take on these battles. She quickly fixed both situations and protected me during the process. Also, during that year, when I was most uncertain as to whether the UA was a good place for me, Dr. Gerken reminded me that I was at the UA by choice (i.e. I am a free agent); I could leave if things did not get better. In all the ways that the department and the university could have felt oppressive or stifling, her feedback and advice consistently created psychological and intellectual space for me in the department, the university, and the academy as a whole.

As I reflect back on our interactions, what impressed me most is that she took the time to learn more about me (i.e. my personal and professional motivation and goals) and the situation before offering advice, and she never forced me into some preconceived model or ideal of how a junior faculty member should act or behave. In fact, one of my favorite moments was when I asked her whether accepting an invitation to give a talk that required traveling for several days was a good idea. She replied, "Well, if you were a typical junior faculty member, I would say no, but since you aren't and since you seem to be able to handle many things at once, I think you should do it." I cannot express the sense of trust and relief this comment provided.

Another key aspect of the mentoring relationship that served to bridge racial-ethnic differences is that we spent a lot of time talking and developing shared language about what constitutes good mentoring. For example, a topic we discussed often was "conservative mentoring," which we referred to as mentoring junior faculty of color and junior women to be professionally cautious—to engage in safe research, to say no to invitations to give talks and to write chapters, and to say no to service opportunities. Dr. Gerken worried that conservative mentoring encourages more traditional pathways for success and potentially discourages junior faculty from taking risks and from exploring alternative professional pathways that may fast track their careers. Hence, she encouraged me to think creatively about my professional and personal goals. These discussions gave me confidence that Dr. Gerken understood my identity-related experiences and ultimately, when difficult situations arose, they provided the shared language or common ground needed to resolve the situation.

As my career progressed and the sense of struggling increased, Dr. Gerken importantly and frequently reminded me that being a junior faculty member was a small part of my career and she encouraged me to stay focused on my larger professional goals. This does not mean that she encouraged me to disregard promotion and tenure expectations or to say "yes" to everything, but rather that we discussed various activities related to my long-term goals and whether they would propel my career or simply become another line on my CV. If we determined there was a clear benefit, then I participated, and if the benefit was ambiguous or small, then I did not participate. The key, however, is that the discussion did not center only on promotion and tenure expectations, but rather was considered in the context of my whole career and my professional goals. As a result, I learned to think about my own professional and personal life in innovative ways that made the university feel like a place where I could grow rather than a place where I felt stifled.

Finally, both mentors served the important role of providing "reality checks" for me. I trusted them, so at the lowest points in my career I asked for an honest opinion as to whether I was struggling as a junior faculty member. Both mentors seemed to recognize the important role these assessments played in my personal and professional well-being. Both took them seriously, but generally responded quite differently. Dr. Markus empathized with my feelings, but also engaged in a useful reframing of the situation. She asked questions about the situation and the people involved in the situation so that I could see the many factors influencing how I was feeling. She also reminded me of the many barriers I had overcome in my life, emphasizing that "struggler" was not part of my identity. Dr. Gerken engaged in similar activities, but importantly helped to normalize my experience. She related my feelings and my experiences to past junior faculty whom she had successfully mentored for tenure and she conveyed to me a sense of confidence that I too would be successful. While the reality checks did not completely alleviate the feeling of struggling, they often provided enough relief that I could return to being a productive researcher, teacher, and service-contributing member of the university.

Affiliation

Two institutional affiliations also played vital roles in my effort to overcome the struggler phenomenon. The first is the Center for Comparative Studies in Race and Ethnicity (CCSRE) at Stanford University. In graduate school, CCSRE was very influential in my development as a scholar. I participated in faculty networks, attended lunchtime speaker series, and met researchers from around the country. In essence, CCSRE provided examples of "successful academics" who were people of color, who were people like me. When I began to question whether I belonged in the academy or whether I could be successful, I saw my future self in these influential scholars. Their works and their images sustained me during the difficult times, and they led me to believe that success was possible.

The second affiliation is the Future of Minority Studies (FMS) national research project, a multi-institutional consortium of scholars interested in minority identities (e.g., race, gender, disability, and sexuality), educational equity, and social and/or institutional transformations. FMS is the single most important group with which I am currently affiliated. For two or three weekends a year, I share experiences and research interests with faculty and administrators, of various racial-ethnic backgrounds, at all levels of the university, and from a wide variety of disciplines. For two or three weekends a year, I experience a place, often the only place, where I have a sense that I truly belong. I do not have to educate people about my experiences, ignore persistent inequalities, or attend to other people's comfort. At FMS, I am just one among many who share collective motivations for change and who are willing to risk personal success for the sake of community gain. In this space, I feel safe to share ideas and personal experiences and, as a result, I experience tremendous personal and intellectual growth.

When I first recognized the struggler phenomenon, I was sitting in a morning session at the FMS National Conference at the University of Oregon. Professor Linda Martín Alcoff, Professor of Philosophy, Chair of Women's Studies at Syracuse University, and one of the founding members of FMS, was speaking about the effects of the black-white race binary on other racial-ethnic minority groups (i.e. Chicanos and American Indians). I am a big fan of Professor

Alcoff, both as a person and as a scholar. In addition to being a genuinely kind and gracious individual, she epitomizes the type of professor I aspire to be: an outstanding scholar, a dedicated mentor, and an influential contributor to her field, to interdisciplinary studies, and to society. At the conference, listening to Professor Alcoff speak, I found myself wondering whether I could ever achieve her level of success or status, and, in my post-mid-tenure dossier writing state, I found myself doubtful.

What seemed to be driving the comparison and the doubt is that I submitted my mid-tenure review dossier a few days before the conference, and frankly I was still recovering from the anxiety that the process of creating the dossier and thinking about tenure evoked. As I began to question my professorial self yet again, I literally had a "moment" where I stopped thinking about my struggler self and I started thinking about what Professor Alcoff represents for me. First, I contemplated whether my experience might have been different if Professor Alcoff had been a faculty member in my department. I thought about the inspiration I gained from Professor Alcoff's work and from my past discussions with her. I thought of her struggles in the academy, the stories of facing hardships and of developing strategies to cope and to survive. I thought of the times she encouraged my ideas and even applauded my work. I thought about the ways in which I am like her—we share similar research interests (albeit from different disciplines); a desire to work toward educational equity; and commitments to issues of social justice for underserved communities. As these thoughts raced through my head, I realized that the doubts and concern about being viewed as a struggler were not mine alone, but were also shared by other junior faculty of color, as well as senior faculty and university administration, who are influenced by this lack of representation.

I then considered what Professor Alcoff would represent to senior faculty and university administrators. Would I have heeded all those warnings if Professor Alcoff had been a member of my department? I wondered whether Professor Alcoff's "power," so to speak, lies solely in her scholarship and style of mentorship or in a larger representational story being told by her presence, her survival, and her success in the academy. Most of my colleagues believe explicitly that anyone can be successful, but they do not recognize the ways in which their definition of success includes the experiences of some people but not others. This is when I realized that the struggler phenomenon is deeply entrenched in the cultural ideas and practices of the university.

While the number of successful senior faculty of color is slowly increasing, the reality is that the academy has been slow to change.

Old images of "successful academics" persist and young scholars, like me, continue to feel their effects. Until universities actively rethink the prevalent conceptions of success, junior faculty of color will continue to face the struggler phenomenon and the uncertainty of their professional survival. Moreover, although institutes such as FMS and CCSRE and exemplars such as Professor Alcoff represent an excellent step in the right direction, at a more foundational level, successful diversity initiatives will require that departments and universities recognize and value the contributions of faculty of color—as contributors who are essential to educational equity and social justice.

Disidentification

When individuals experience chronic threat in a situation, such as the uncertainty of their potential survival, they may become increasingly less identified with achieving success in the domain.[11] Individuals disidentify because some aspect of the domain, such as the presence of negative stereotypes, poses a threat to their self-worth.[12] While disidentification may sometimes represent a fairly adaptive coping response, it can also lead to decreased motivation and reduced performance.[13]

In my case, I believe that initially disidentification seriously impeded motivation and performance, but eventually served as an adaptive response—it allowed me to disregard the negative stereotypes and to continue to pursue my passions. When I first joined the professoriate, my motivation for research, teaching, and service was not tied to achieving tenure, but rather to helping my community, to mentoring students, and to increasing educational equity in the university. However, after repeated exposure to "struggler" messages about people like me, I became increasingly concerned about my prospects for tenure. By eventually disidentifying with the academy, I was able to return my attention to the issues for which I had a deep sense of commitment and passion.

While I do not advocate this strategy for everyone, there is much to learn from it. First, the phenomenon of disidentification with one's department or university, or even the academy as a whole, may help explain why universities have difficulty retaining junior faculty of color. Moreover, the research suggests that the faculty members who leave are likely to be those who are the most identified with achievement, precisely the individuals the academy should aspire to keep.[14] In order to begin the process of alleviating disidentification, the academy can start by taking stock of public representations—by decreasing negative stereotypes (e.g. women and minorities do not do well here) and increasing positive representations about junior faculty of color. While eliminating stereotypes is not the whole answer, this representational change minimally serves to acknowledge that junior faculty of color belong in the academy, that some frustration and failure is okay, and that, if junior faculty of color stay in the academy, then success may be just around the corner.

SOCIAL SCIENCE RESEARCH: UNPACKING THE STRUGGLER PHENOMENON

While most junior faculty find the tenure process stressful, the struggler phenomenon illustrates barriers that make the experience of being junior faculty and of attaining tenure more difficult for junior faculty of color. As my personal narrative reveals, in graduate school I had an outstanding mentor, a number of stellar successful academics of color as role models, and a wealth of information about how factors such as negative stereotyping and numerical underrepresentation influenced psychological well-being and performance. As a junior faculty member, I won a university-wide award for excellence in undergraduate teaching and an early career award for contributions of my research to society, and yet the struggler phenomenon has been a very real part of my junior faculty experience. What about junior faculty of color who do not have access to the same advantages and/or who do not receive these types of explicit validation? How do they contend with the struggler phenomenon? In this section, I utilize social science research about social identity threat, solo status, and individualism to unpack factors associated with the struggler phenomenon.

Social Identity Threat

Cultural ideas and/or stereotypes about "successful academics" and "junior faculty of color struggling to attain tenure" convey messages to all members of the academy about who and what constitutes an acceptable faculty member, who belongs, and ultimately who deserves tenure. These ideas, whether fueled by institutional bias, low expectations, or the inability to imagine faculty of color as "successful academics," heighten junior faculty of color's concern that they will be unfairly evaluated or that they will be seen through the lens of a negative stereotype. Steele, Spencer, and Aronson refer to the association between social identities and negative stereotypes as social identity threats, and their research, as well as the work of other stereotype threat researchers, suggests that these concerns may weigh heavily on the professional performance and psychological well-being of junior faculty of color.[15]

While the research and service of faculty of color are often undervalued or marginalized in the academy, the situation is further complicated by the fact that junior faculty of color chronically face the prospect of devaluation (i.e. that they will be unfairly evaluated or that they will be seen through the lens of a negative stereotype).[16] The social identity threat literature, also known as stereotype threat, suggests that concerns about confirming the negative stereotypes contribute to the underperformance of junior faculty of color.[17]

For example, to illustrate how stereotype threat works, Steele and Aronson tapped into the stereotype that African Americans are intellectually inferior to European Americans. They gave European American and African American students a difficult verbal test from the Graduate Record Examination (GRE).[18] They told half of the students that the test measured intelligence and the other half of the students that the test was a problem-solving exercise that was not indicative of ability. Steele and Aronson found that European American students performed equally well regardless of how the test was described, but African American students performed worse when the test was described as an intelligence test than when the test was described as a problem-solving exercise. They argued that when African American students were told that the test measured intelligence, the information elicited concerns about not performing well and thus confirming the stereotype, and these concerns ultimately interfered with performance. Steele and Aronson concluded that negative stereotypes raise the specter of a racially stereotypic interpretation of performance.

In the academy, negative stereotypes about the underperformance of junior faculty of color may evoke stereotype threat. When junior faculty of color experience setbacks, even normal setbacks, such as a rejected manuscript, or receive performance advice, such as encouragement to publish faster, the impact is likely to be amplified by the stereotype. The setback becomes "evidence" that the individuals are either struggling or could be seen as struggling if they do not publish faster. As the evaluative pressures of tenure increase (i.e. they receive annual or mid-tenure reviews or tenure evaluation nears), the concern about performance and about potentially confirming the negative stereotype that they will struggle also increases. Over time, the concern may influence the ability to concentrate and ultimately to be productive researchers, which inadvertently serves to confirm the stereotype.[19]

The power of stereotype threat is that knowing about the effect is not enough to alleviate the effect. For example, in graduate school, one of my African American colleagues studied stereotype threat. He is exceptionally bright, but he struggled with showing people his writing. The issue was not that his writing ability was below average or even average, but rather that writing was a skill with which minority students were thought to need a lot of help. As a highly identified, high-achieving graduate student, he feared being seen as one of those students. While he was acutely aware that he was experiencing stereotype threat, in general, he was paralyzed by the fear and was never able to overcome the threat. He eventually left the professoriate.

I was a few years behind my friend in graduate school. Watching him struggle with stereotype threat made me fearful of my own future in the academy. Fortunately, my friend's experience prompted me to talk with my graduate advisor. I explained to her my fears about writing and about my future in the academy. Without speaking, Dr. Markus, who is a highly acclaimed social psychologist, pulled out the first draft of a paper she was writing. As I began to read through the pages of relatively incomprehensible text, I felt a sense of confusion, but also a sense of relief. Dr. Markus told me that when she writes she starts by getting the ideas on paper. Every paper, as I so painfully learned, is ten drafts from the final product. On this occasion, Dr. Markus invoked two strategies that have since been shown to reduce stereotype threat. First, she reframed the task as being a process and a learned skill rather than a measure of innate ability and, second, she emphasized that writing, for all people, improves with effort.[20]

Shortly after this discussion, I began sharing "first drafts" with Dr. Markus. She was careful not to critique my writing, but would rather focus on creating the next draft. Over the years, we spent numerous hours together talking through ideas, outlining papers, and writing and rewriting sentence after sentence. The goal was always the next draft and the standards for the "final draft" were always high. One reason this strategy worked is that I trusted Dr. Markus, but she also became a positive role model for writing and she helped me believe I could achieve the high standard she set.[21] These are also important contingencies for alleviating stereotype threat.

When I became a junior faculty member, the messages I heard about women and minorities struggling to attain tenure made me once again fearful about my future in the academy. The high bar of excellence had been raised and the specter of low expectancies for people like me was once again salient. The stereotype threat research reveals that under the right set of social identity threat contingencies, stereotype threat can negatively affect any person (i.e. even European American males).[22] This is not because individuals choose to "believe" in the stereotype or because they are weak individuals, but rather because the threat of the situation—the concern that others may view you through a negative stereotype—can powerfully influence the individual's behavior.

In fact, beliefs about the stereotypes do not protect individuals from the effects of the stereotypes. For example, regardless of whether junior faculty members of color consciously acknowledge or disavow the "struggler" stereotype, they will still be affected by the stereotype.[23] As Claude Steele noted, the power of stereotypes is that they exist in the world, in the ideas, practices, and institutions in which people participate, not just in the minds of stereotyped individuals.[24] In other words, although people can try to "choose" to disregard a stereotype, this choice will not eliminate the experience of stereotype threat unless those individuals also have the power to decide what other people believe or to change the cultural ideas and stereotypes that are present in the social world.

Finally, stereotypes do not impact the mentally "weak"; they impact the highly committed. Junior faculty of color who are the most highly identified with being productive scholars and with achieving tenure are especially vulnerable to stereotype threat.[25] Hence, the current culture of the academy may be undermining the performance of highly dedicated individuals who, without all the evaluative pressures and concerns about their ability, would be influential contributors to the academy and to society.[26]

Solo Status

Social identity threats can also be amplified if junior faculty of color experience what psychologists call "solo status" in the department. *Solo status* refers to being the only member of one's social identity group (i.e. gender, social class, or racial group) in an otherwise homogeneous group.[27] Junior faculty of color who are the only members of their race (i.e. race solos) in the department are likely to experience heightened visibility, stereotypical role encapsulation, and contrast effects.[28] These factors, which I review in detail later, can render ineffective the academy's efforts to be more inclusive.

The experience of heightened visibility can be evoked by a variety of situations. For example, if the faculty is largely homogeneous (i.e. majority European American and/or male), then the composition may remind junior faculty of color of their unique status. Moreover, in homogeneous groups, individuals are less likely to have experience with diverse individuals and thus the presence of a race solo is even more noticeable to individuals within the dominant group. The solo situation also creates higher stakes for both majority and solo faculty members.[29] An awkward or difficult interaction, for example, may heighten the anxieties of the majority faculty members about being seen as racist, whereas the solo faculty members may carry the burden of ensuring that majority faculty members feel comfortable.[30] Research suggests that solo faculty are likely to be negatively evaluated if they do not ensure the comfort of majority faculty.[31]

Another consequence of solo status is that junior faculty of color are often assigned stereotypic roles, what is referred to as stereotypic role encapsulation.[32] Junior faculty of color, for example, get asked to serve on various diversity-related committees, whether they express interest in diversity issues or not. The placement of junior faculty of color on these committees is often viewed as necessary for ensuring a more inclusive campus. The issue, with respect to stereotypic role encapsulation, however, is that these committees are often the only committees for which junior faculty of color are considered. They are not being considered for roles that may best fit their personalities, interests, or future prospects.[33]

A third consequence of solo status is a contrast effect, where the in-group is seen as more similar because of the perceived contrast with the out-group.[34] Contrast effects occur because people generally, by default, unconsciously judge other people according to their most salient features or identities in a given situation.[35] For example, a faculty member of color in a group of white faculty members is likely to be unconsciously judged according to race, whereas a junior faculty member in a group of senior faculty is likely to be unconsciously judged according to faculty status. The contrast effect may inadvertently lead majority faculty to assess the opinions, behaviors, and personality traits of junior faculty of color as more extreme and as related to their most salient features, such as their race or faculty status.[36] As a result, majority faculty may place more social and physical distance between themselves and the junior faculty member of color.[37] For example, they may not invite junior faculty of color to social gatherings or to other informal opportunities for networking.[38] In sum, contrast effects lead to evaluations of junior faculty of color based on stereotypic expectancies about attitudes and behaviors.

In the academy, solo status for junior faculty of color is further complicated by their low status (i.e. being junior faculty). Take a simple request made by a senior faculty member to a junior faculty member of color to sit on a committee. While the request may be framed as a choice, junior faculty of color may find the request to be more of a demand.[39] The heightened visibility of junior faculty of color creates a situation in which declining the request will be remembered. At a later date, when the senior faculty member is asked to vote on the tenure or promotion of junior faculty of color, their response may be inadvertently influenced by the past interaction. In other words, while the senior faculty member's perception is that the junior faculty member of color was given a choice, the situation of solo status usurps the junior faculty member's experience of choice.

In fact, the combination of solo status and low status (i.e. being a junior faculty member) has a number of negative consequences for performance expectations and task performance.[40] In a study of job performance ratings by gender and race compositions in the workplace, Sackett, Dubois, and Noe revealed that when men outnumber women in the workplace, women's job performance ratings were consistently lower than men's; and when European Americans outnumber African Americans, African Americans' job performance ratings were consistently lower than those of European Americans.[41] However, when women constituted more than 50 percent of the employees, their ratings were equal to the men's evaluations and when African Americans composed 90 percent of the employees their ratings were equal to those of European Americans.

Similarly, in a study of psychology professors with solo status, Niemann and Dovidio found that faculty whose groups were relatively more underrepresented and more negatively stereotyped reported greater difficulties in their professional environments.[42] In particular, black professors with solo status reported greater difficulties than did Latino professors, and Latino professors reported greater difficulties than did professors of Asian descent. The more underrepresented the faculty member's group, the more vulnerable its members were to the harmful effects of solo status. The absence of American Indian professors in the study is a result of the low numbers of American Indian faculty in psychology, but it also speaks to the striking invisibility of this group in the academy. When individuals observe that their group is underrepresented, they are less likely to enter that setting and they are more likely to believe that they would not perform well if they were to enter the setting.[43]

In summary, the underrepresentation of faculty of color is yet another reason why universities have difficulty recruiting and retaining junior faculty of color. In situations of solo status, junior faculty of color are more likely to experience heightened visibility, stereotypic role encapsulation, and contrast effects, and these factors are likely to influence productivity and job performance. This is true whether or not social identity threat is present. Sekaquaptewa and Thompson, for example, found that the effects of solo status and stereotype threat on performance were both distinctive and additive.[44] Women in the solo status but no-stereotype threat condition performed more poorly than women in the nonsolo no-stereotype threat (control) condition. Eliminating stereotype threat did not eliminate the effect of solo status on women's performance. Similarly, eliminating solo status did not eliminate the effects of stereotype threat for women. In the nonsolo stereotype threat condition women performed more poorly than women in the nonsolo no-stereotype threat condition. Finally, revealing the additive effects, women in the solo status *and* stereotype threat condition performed significantly worse than women in the solo status only condition or women in the stereotype threat only condition.

While solo status and stereotype threat are distinct experiences, many of the strategies for alleviating stereotype threat can also be used to alleviate solo status. Increasing the number of faculty of color, for example, may enhance productivity and job performance.[45] The presence of same-race faculty may also improve performance and decrease concerns about stereotyping.[46] This is particularly true if the same-race faculty member is among those evaluating the junior faculty member of color's performance. Similarly, same group (race or gender) mentors increase feelings of belongingness, decrease feelings of distinctiveness, and decrease concerns about stereotyping.[47] In many cases, creating the same group mentor effect merely requires finding common in-group identities between mentor and mentee.[48] As a woman of color, I found this sense of common in-group identities by selecting race-savvy white women. We shared the common identity as women, but their knowledge about race issues provided a sense that I could trust them to understand and validate my experiences as a faculty member of color. Moreover, their senior status provided me evidence that I belonged and that I could be successful in the academy.[49]

Individualism and Meritocracy

This chapter began with a discussion of the ways in which the culture of the academy influences pervasive ideas about what constitutes "successful academics." I want to return briefly to these ideas to discuss the influence they have on collective understandings of social identity threat and solo status. In President Tilghman's speech, for example, she stated that her own resistance to the stereotypic view that women were not meant to do science was prompted, among other things, by an "absolute inability to recognize reality" and by a refusal to allow herself to become a victim:

> It has been my experience that many successful women in science simply fail to perceive that there are obstacles in their path. They are able to go through life with metaphoric blinders on—not that they would deny that there are forces working against the progress of women, but rather that they refuse to acknowledge that those forces apply to them. A blunt way to describe such women is to say that they refuse to allow themselves to become victims.

In this statement, President Tilghman taps into a powerful cultural belief about individualism and meritocracy. First, with respect to individualism, she locates the source of success (i.e. motivation) "within" the successful women and she highlights the fact that these successful women refuse to allow outside forces, such as stereotype threat or solo status, to impede their success. In other words, she implies that the women can simply "choose" not to let the stereotypic views influence them. The individual is the sole arbiter of success and failure and the situational factors or circumstances (i.e. personal history, family background, stereotype threat, and solo status) are relegated to the periphery.[50] The message conveys to young women that if they work hard, ignore what other people say or do, and persist through difficult situations, then success can be attained.

President Tilghman's message, in this light, is the culturally correct message. Individuals should not allow other people or situations to push them out of a particular field of study or an area of achievement. Unfortunately, the message also downplays the powerful role that stereotypes or cultural ideas about race and gender can play in performance. From this perspective, social identity threat and solo status are merely impositions that successful individuals can transcend, rather than major obstacles that unfairly undermine motivation and performance. If women or junior faculty of color fail, then the implication is that they "chose" to fail (i.e. that they are personally responsible). In other words, they failed because they decided to buy into the stereotypes and to allow themselves to become victims.

In the United States, for example, people believe that merit should be based on demonstrated talent and ability, rather than on wealth, position, or social status.[51] These ideas stem from the philosophical and political writings of John Locke, but they are deeply entrenched in American cultural beliefs.[52] Locke argued that society will inevitably be stratified, but that the stratification will be considered just or fair if it is based on merit, rather than on inheritance. Locke's theory applied to kings and governments of nobles because they obtained high status based on birthrights. Today Americans continue to believe that individuals should be judged by what they do, not by their background, but now, ironically, what people do is often influenced by the privileges associated with their background.[53] By continuing to disregard the privilege associated with backgrounds, Americans ignore distinctions based on race or gender.

Unfortunately, race and gender have serious consequences in America. Hiring the "best" or the "most qualified" individual feels like the right practice, but, in actuality, the practice only ensures equity if everyone is presumed to begin with the same resources and opportunities to succeed and if those in positions of selecting the "best" are aware of how their individual biases influence their decisions, particularly as their biases relate to social identities.[54] One advantage of privileging individual achievements over social identities, for example, is that it affords those who succeed the belief that their own achievement is a product of good choices, hard work, and ability and that those who do not succeed are not working hard or that they lack the ability to be successful. By locating the source of motivation within the individual and by expecting individuals to transcend their situations, individualism and meritocracy become the justification for denying the barriers underprivileged individuals face and ultimately for deeming them as "falling short." In the academy, these particular notions of individualism and meritocracy disguise the difficulties junior faculty of color face.

Most university administrators and senior faculty acknowledge that situations can influence performance, but they struggle with how to "fairly" assess the impact of the social barriers individuals face. Should individuals be "punished" because their parents had the resources to send them to the best schools, provide them with tutors and coaches, and guide them in the right direction? Conversely, should individuals be disadvantaged because their parents did not have such resources? Or, should they be "bumped up" or viewed more positively because they managed to reach a high level of excellence, despite these obstacles (e.g. social identity threats, solo status)? In other words, weighing situational factors runs the risk of undermining the perceived rights of individuals and the absolute assessment of their merit (i.e. comparing the number of publications, teaching evaluations, and number of committees), but I contend here that not doing so also undermines these rights and assessments of merit.

University administrators and senior faculty, for example, often sidestep this issue by "helping" junior faculty of color avoid these obstacles during the pre-tenure years. They encourage them to reject the stereotypes, to rise above the solo status, and to privilege scholarship over service and teaching. Ultimately, junior faculty of color need only heed their advice and "choose" to say, "No." Given the "help," junior faculty of color are then viewed as having had equal opportunities for success and thus are expected to measure up to their white junior faculty peers. This perspective blinds university administrators and senior faculty from seeing the systemic inequalities (e.g. social identity threat, solo status) that exist in the university and leave junior faculty of color vulnerable to a set of conditional terms of inclusion that inherently disadvantages them.[55]

For example, universities want to evaluate junior faculty of color as individuals, but they ignore the fact that in the absence of sufficient numbers of senior faculty of color, universities need junior faculty of color to be members of their groups. They need them to help meet the needs of an increasingly diverse campus—to "represent" diversity and diversity issues on committees, to mentor students of color, to educate majority students, faculty, and administrators, and to help accomplish the mission of the academy (i.e. to provide a high-quality education to all students). Instead, universities ignore this reality and insist that in order to stay in the academy (i.e. to attain tenure), junior faculty of color must ignore the service and mentoring needs of these communities for six years or longer, depending upon the particular university, while they prove that they can conduct tenure-worthy research. In other words, universities need junior faculty of color to fulfill these obligations; they just need them to fulfill them under the guise of "choice."

Unfortunately, junior faculty of color do not get to live out this individualism or this illusion of choice between helping needy communities and being denied tenure for doing too much service. The reality is that this choice is not a choice at all—junior faculty of color are forced into a quintessential lose-lose situation. On the one hand, junior faculty of color experience the inequality in the academy and in society at large, and they recognize the subsequent needs of these communities (e.g. lack of mentorship, faculty representation, community connection). Research has shown that most junior faculty of color justify staying in the academy because they are helping (i.e. mentoring, serving as role models) students and communities of color to face these inequalities and because they are working toward institutional change, rather than just accepting the university in its current form.[56] Asking junior faculty of color to turn their back on their communities is inherently unfair—university administrators and senior faculty are asking them to ignore these communities in order to protect their own future, when they "might" be able to help them and still get tenure. This reality produces what W. E. B. Dubois termed a "double consciousness."[57] They realize that they should privilege their own individual success, but they realize that at the end of the day they will not be judged as individuals, but as a member of a group. In this sense, helping their communities, rather than protecting themselves and inadvertently contributing to the existing inequalities, is not a choice. It is a personal, professional, and political necessity.

On the other hand, even if junior faculty of color could turn away their group(s), the consequences would be dire. First, they need their group affiliation. Everyone needs group affiliations, but when you are a member of the minority, you feel this need even more. White faculty are constantly affiliated with their group; in fact, they are helped by their group affiliation in a variety of ways.[58] Being in the majority, however, means that they do not have to recognize their group affiliation or its influences on their motivations. There are majority group individuals in the academy who freely choose to dedicate their lives to university service and consequently white faculty are free to choose scholarship over service. Faculty of color, however, do not have the same choice. They are underrepresented in most domains of the academy and, as a result, for the academy to change (i.e. to become more inclusive) they must conduct both service and research. Second, they need their group affiliations in order to enhance feelings of belonging and to envision future possibilities for success. Asking junior faculty of color to put this aside is akin to asking them to actively reify the system that, in many respects, excludes them. Such a request undermines progress toward a more diverse and equitable university, and contributes to a desire, on the part of faculty of color, to leave.

STRATEGIES FOR CHANGE

In order to retain junior faculty of color and to encourage their promotion and tenure, the cultural ideas and stereotypes about "successful academics" and "junior faculty of color as strugglers" must be reexamined. The problem is not that junior faculty of color are not "successful academics," but rather that universities as they currently operate (i.e. with their particular definitions of success) are unable to see them as such. Until the prevalent beliefs about junior faculty of color change and the current images of successful academics—which produce guidelines for promotion and tenure—include the motivations, experiences, and important contributions of junior faculty of color, then efforts to diversify the academy and to help the academy reach its full potential will also fall short.

In this essay I have focused on the professional and psychological toll exacted by the struggler phenomenon, the strategies that helped me to survive the experience, and the research that is relevant to the phenomenon. In the final section, I briefly outline strategies that junior faculty of color, university administrators, and senior faculty can utilize to reduce the struggler phenomenon.

Strategies for Junior Faculty of Color

I want to be clear here that the onus for change does not lie with junior faculty of color. The struggler phenomenon does not exist in the minds of junior faculty of color, but rather in the cultural ideas and practices of universities and in the minds of university administrators and senior faculty. Unfortunately, until the academy changes, junior faculty of color will need to be aware of the situations that can influence their professional and psychological well-being and they need to situate themselves in the best place possible. I offer these strategies merely as a place to begin.

1. *Recognize the struggler language*: The first step to changing the situation is to recognize the particular messages university administrators and senior faculty convey about junior faculty of color (e.g. junior faculty of color do not do well here, junior faculty of color struggle to attain tenure, junior faculty of color engage in too much service and not enough research). By recognizing the language, you can help your colleagues to recognize it and you can help alleviate the added pressure created by the struggler language.
2. *Be cognizant of social identity threat and solo status*: Being a race or gender solo can have a variety of psychological and professional consequences; it can undermine motivation, performance, and well-being, and it can consume time and energy for scholarship (e.g. you can get asked to be on every committee, to mentor every student of color, and to champion diversity initiatives). Educate yourself and your colleagues (if they are willing) about these important situational factors. You can protect yourself and help your colleagues protect you. Choose service activities that will positively impact your career and your future aspirations for the academy.

3. *Concretize your motives*: Make a list of the reasons why you chose to be a professor. When times are difficult, the list will keep you focused on your goals and will prevent your motives from being subverted by external pressures.
4. *Acknowledge that everything you need to know about being a professor is not taught in graduate school*: The skills that made you a successful PhD candidate are not the same skills that will make you a successful professor. Ask colleagues to share best practices, so you can adapt to the culture of the university and so you can do your job more efficiently. Seeking out advice will help you develop new relationships with colleagues and move through the learning process more quickly.
5. *Cautiously find mentors*: Take time to get to know (i.e. interview) faculty until you find a person who is savvy about the issues (e.g. race, culture, sexuality, gender, etc.) that are important to you and who you can trust to provide an honest opinion (i.e. a reality check) when you need and/or want one.
6. *Find relevant professional affiliations*: Figure out what kinds of affiliations will help you grow professionally and will sustain and reenergize you personally. Note, if you are aware of the affiliation prior to negotiating your junior faculty contract, you can ask your prospective employer to fund travel to the meeting or conference.
7. *Be creative about the tenure trio (i.e. research, teaching, and service)*: View the trio as broad categories and try to use them to your advantage. For example, you can save time by combining your research, teaching, and service goals and, if you are creative, you can spend more time on the activities that are important to you.

Strategies for University Administrators and Senior Faculty

President Tilghman stated that universities "stand at the end of a long and imperfectly constructed pipeline...yet this does not excuse us from fixing leaks—and there are many—in the section of the pipeline that we do control." The onus for real change lies with university administrators and senior faculty. These two constituents set the conditions necessary for individuals to stay in the academy. Given this, here are a few suggestions for change:

1. *Make visible the history of diversity efforts for the university*: Most university administrators and senior faculty acknowledge that

diversity efforts are necessary to better accommodate the increasing diversity of the campus communities, but the history of these efforts is often unarticulated. One problem with not articulating this history is that it fosters a sense that the university is making repeated concessions in order to increase the number of brown faces on campus. This sense breeds resentment and hostility toward efforts to enhance equity in the university. The alternative, articulating the history of diversity efforts on campus, allows members of the community to link diversity efforts with strategic, planned change. Moreover, it reminds underrepresented members of the community, in particular junior faculty of color, that some frustration is inevitable, but that greater equity is possible with persistence and effort.

2. *Reexamine the underlying culture of the university*: Universities must reexamine the pervasive ideas about academic excellence and diversity efforts. First, universities need to identify the objectives it hopes to achieve through diversity efforts. Is the goal to increase the representation on campus or to foster an academic climate that benefits the entire campus community? Second, what is the current state of diversity efforts with respect to the mission of the university? Do diversity and excellence go hand-in-hand or do ideas about excellence appear to be in conflict with ideas regarding increasing diversity? Third, what are the pervasive ideas about success? Do images of "successful academics" encourage expansive forms of knowledge generation, innovative pedagogical approaches, and new pathways for mentorship and success, or do they advocate traditional and often ineffective (or outdated?) models of success? Fourth, what language is used to discuss those individuals historically underrepresented at the university?

3. *Evaluate the campus environment—be vigilant for change*: The demographics of the academy have changed quickly in recent decades, but universities have struggled to adapt to these changes. One reason universities struggle to adapt is that the reasons for change are seemingly at odds with the history of the university (e.g. the alumni). Research has demonstrated tremendous benefits for increasing diversity within the university.[59] The objective is not to include diverse individuals and to ask them to assimilate, but rather, it is to include all groups (including majority) and ask all members of the campus to learn from one another—to create a more inclusive learning environment. University administrators and senior faculty must begin by examining their roles in maintaining the status quo (i.e. the way things were), but they must

also open lines of communication from low-status, historically silenced groups (i.e. junior faculty of color). By being vigilant in evaluating the campus environment, universities can grow their potential.

4. *Reexamine promotion and tenure guidelines*: In order to accommodate the increasing diversity of the workforce, many businesses have turned to a portfolio model to examine individual contributions to the company. The idea is that different people bring different skills to the workplace and that achieving excellence requires recruiting and retaining all types of excellence. If universities, for example, only tenure great researchers, then classrooms may be filled with sub-par teachers, thereby reducing the quality of the education provided to students. Utilizing a portfolio practice rather than the strict outlines of the dossier may help grow both diversity and excellence.

5. *Create a reward structure that supports the mission of the university*: At most universities, the most prestigious awards are typically bestowed upon productive, quality researchers (e.g. Regents Professors, Endowed Chairs), with no comparable awards for teaching or service (often, $500 and a plaque). Admittedly, universities, especially public universities, need grant funds, which are usually tied to research, to be financially viable, but universities also need faculty committed to teaching and service to achieve the larger mission of the university (i.e. provide high-quality education to students). This is particularly true with respect to diversity initiatives. If universities want to increase the diverse representation and to meet the needs of a more diverse campus community, then universities must pay tribute to those individuals who champion these causes—they must see them as excellent.

6. *Develop formal, recognized mentoring programs*: Successfully mentoring junior faculty, in particular junior faculty of color, requires an awareness of the particular issues they face. Specifically, it entails confronting and changing the cultural ideas and stereotypes that persist within the university, and necessitates giving voice to low-status faculty (i.e. communication shatters the myths or stereotypes that silence creates). The alternatives—no mentoring, informal mentoring, or formal mentoring without formal training or recognition—are insufficient because they leave quality mentoring to luck.[60] By formally training and rewarding senior faculty, universities can provide junior and senior faculty clear expectations for the mentoring relationship. These shared expectations ease concerns, on the part of junior faculty, that they are

burdening senior faculty, and heighten the expectation and the belief that all junior faculty belong and can be successful at the university.
7. *Develop interdisciplinary centers or institutes that provide faculty of color professional, personal, and intellectual space within the university*: Whether the task is to hold the university responsible for educational equity, provide role models and mentors, or foster research to benefit communities of color, these centers take seriously the personal and professional lives of people of color. They are essential for the personal and professional development of junior faculty of color, but more importantly, they contribute to the recruitment and retention of these individuals. They provide a space where individuals, across disciplines, can come together—this is especially important if the numbers of faculty of color are relatively small. These spaces serve to alleviate feelings of social identity threat, solo status, and alienation.

NOTES

1. I want to thank Satya Mohanty, LouAnn Gerken, Hazel Rose Markus, Ernesto Javier Martínez, Lee Ryan, Nicole Stephens, Sarah Murnen, Dana Mastro, and Karen Francis Begay for reading earlier drafts of the chapter. Finally, thanks to Karshannon Gene and Alem Tecle for helping with references.
2. Shirley M. Tilghman, "Changing the Demographics: Recruiting, Retaining, and Advancing Women Scientists in Academia," *Earth Institute ADVANCE Program*, Columbia University, March 24, 2005.
3. Roy Goodwin D'Andrade, "The Cultural Part of Cognition," *Cognitive Science* 5.3 (1981): 179–195; D'Andrade, "Moral Models in Anthropology," *Current Anthropology* 36.3 (1995): 399–408; Dorothy C. Holland and Naomi Quinn, *Cultural Models in Language and Thought* (Cambridge [Cambridgeshire]; New York: Cambridge University Press, 1987); Bradd Shore, *Culture in Mind: Cognition, Culture, and the Problem of Meaning* (New York: Oxford University Press, 1996); Bradd Shore, *What Culture Means, How Culture Means* (Worcester, MA: Clark University Press, 1998); Dan Sperber, *On Anthropological Knowledge: Three Essays* (Cambridge [Cambridgeshire]; New York: Cambridge University Press, 1985).
4. Alan Page Fiske, Shinobu Kitayama, Hazel Rose Markus, and Richard Nisbett, "The Cultural Matrix of Social Psychology," in *The Handbook of Social Psychology*, ed. Daniel Todd Gilbert, Susan T. Fiske, and Gardner Lindzey, 4th ed. (Boston; New York: McGraw-Hill; Distributed exclusively by Oxford University Press, 1998), pp. 915–981; Li Jin, "U.S and Chinese Cultural Beliefs about Learning," *Journal of Educational Psychology* 95.2

(2003): 258–267; Hazel Rose Markus, Patricia R. Mullally, and Shinobu Kitayama, "Selfways: Diversity in Modes of Cultural Participation," in *The Conceptual Self in Context: Culture, Experience, Self-Understanding*, ed. Ulric Neisser and David A. Jopling (Cambridge, UK; New York: Cambridge University Press, 1997), pp. 13–61; Michael Tomasello, *The Cultural Origins of Human Cognition* (Cambridge, MA: Harvard University Press, 1999).

5. Victoria C. Plaut, "Models of Success in the Academy," in *Engaging Our Faculties: Junior Faculty of Color and Senior University Administrators on Diversity and Excellence in Higher Education*, ed. Stephanie A. Fryberg and Ernesto Javier Martínez (manuscript in progress).

6. Stephanie A. Fryberg and Hazel Rose Markus, "Cultural Models of Education in American Indian, Asian American and European American Contexts," *Social Psychology of Education* 10.2 (2007): 213–246; Alex Kozulin, *Psychological Tools: A Sociocultural Approach to Education* (Cambridge: Harvard University Press, 1998).

7. Anna M. Agathangelou and Lily H. M. Ling, "An Unten(Ur)Able Position: The Politics of Teaching for Women of Color in the U.S.," *International Feminist Journal of Politics* 4.3 (2002): 368–398; Sylvia Hurtado, "The Campus Racial Climate: Contexts of Conflict," *The Journal of Higher Education* 63.5 (1992): 539–569; Sylvia Hurtdao, Jeffrey F. Milem, Alma R. Clayton-Pedersen, and Walter R. Allen, "Enhancing Campus Climates for Racial/Ethnic Diversity: Educational Policy and Practice," *The Review of Higher Education* 21.3 (1998): 279–302.

8. Fryberg and Markus, "Cultural Models of Education in American Indian, Asian American and European American Contexts," 213–246.

9. David Nussbaum and Claude M. Steele, "Situational Disengagement and Persistence in the Face of Adversity," *Journal of Experimental Social Psychology* 43.1 (2007): 127–134; Jessi L. Amith, Carol Sansone, and Paul H. White, "The Stereotyped Task Engagement Process: The Role of Interest and Achievement Motivation," *Journal of Educational Psychology* 99.1 (2007): 99–114; Claude M. Steele, "A Threat in the Air: How Stereotypes Shape Intellectual Identity and Performance," *American Psychologist* 52.6 (1997): 613–629.

10. Holly Angelique, Ken Kyle, and Ed Taylor, "Mentors and Muses: New Strategies for Academic Success," *Innovative Higher Education* 26.3 (2002): 195–209; Peg Boyle and Bob Boice, "Systematic Mentoring for New Faculty Teachers and Graduate Teaching Assistants," *Innovative Higher Education* 22.3 (1998): 157–179; Jeffrey A. Morzinski, Sabina Diehr, David J. Bower, and Deborah E. Simpson, "A Descriptive, Cross-Sectional Study of Formal Mentoring for Faculty," *Family Medicine* 28.6 (1996): 434–438; Deborah Olsen and Lizabeth A. Crawford, "A Five-Year Study of Junior Faculty Expectations about Their Work," *The Review of Higher Education* 22.1 (1998): 39–54; Paul Schrodt, Carol Stringer Cawyer, and Renee Sanders, "An Examination of Academic

Mentoring Behaviors and New Faculty Members Satisfaction with Socialization and Tenure and Promotion Processes," *Communication Education* 52.1 (2003): 17–30.
11. Claude M. Steele, Steven J. Spencer, and Joshua Aronson, "Contending with Group Image: The Psychology of Stereotype and Social Identity Threat," *Advances in Experimental Social Psychology* 34 (2002): 379–440.
12. Jennifer Crocker, Brenda Major, and Claude M. Steele, "Social Stigma," in *The Handbook of Social Psychology*, ed. Daniel Todd Gilbert, Susan T. Fiske, and Gardner Lindzey, 4th ed. (Boston; New York: McGraw-Hill; Distributed exclusively by Oxford University Press, 1998), pp. 504–553; Brenda Major and Toni Schmader, "Coping with Stigma through Psychological Disengagement," in *Prejudice: The Target's Perspective*, ed. Janet K. Swim and Charles Stangor (San Diego, CA: Academic Press, 1998), pp. 219–241.
13. Nussbaum and Steele, "Situational Disengagement and Persistence in the Face of Adversity"; Kevin O. Cokley, "Ethnicity, Gender and Academic Self-Concept: A Preliminary Examination of Academic Disidentification and Implications for Psychologists," *Cultural Diversity and Ethnic Minority Psychology* 8.4 (2002): 378–388; Brenda Major, Steven Spencer, Toni Schmader, Connie Wolfe, and Jennifer Crocker, "Coping with Negative Stereotypes about Intellectual Performance: The Role of Psychological Disengagement," *Personality and Social Psychology Bulletin* 24.1 (1998): 34–50.
14. Jason Osborne and Christopher Walker, "Stereotype Threat, Identification with Academics, and Withdrawal from School: Why the Most Successful Students of Colour Might Be Most Likely to Withdraw," *Educational Psychology* 26.4 (2006): 563–577.
15. Steele, Spencer, and Aronson, "Contending with Group Image."
16. Anthony Lising Antonio, "Faculty of Color Reconsidered: Reassessing Contributions to Scholarship," *The Journal of Higher Education* 73.5 (2002): 582–602; Mary Jo Tippeconnic Fox, "American Indian Women in Academia: The Joys and Challenges," *Journal about Women in Higher Education* 1 (2008): 202–221; John W. Tippeconnic and Smokey McKinney, "Native Faculty: Scholarship and Development," in *The Renaissance of American Indian Higher Education: Capturing the Dream*, ed. K. P. Maenette, AhNee-Benham, and Wayne J. Stein (Mahwah, NJ: Lawrence Erlbaum, 2003), pp. 241–296; Caroline Sotello Viernes Turner and Samuel L. Myers, *Faculty of Color in Academe: Bittersweet Success* (Boston, MA: Allyn and Bacon, 2000); Octavio Villalpando and Dolores Delgado Bernal, "A Critical Race Theory Analysis of Barriers that Impede Success of Faculty of Color," in *The Racial Crisis in American Higher Education: Continuing Challenges for the Twenty-First Century*, ed. William A. Smith, Philip G. Altbach, and Kofi Lomotey (Albany, NY: State University of New York Press, 2002), pp. 243–270.

17. Claude M. Steele and Joshua Aronson, "Stereotype Threat and the Intellectual Test Performance of African Americans," *Journal of Personality and Social Psychology* 69.5 (1995): 797.
18. Ibid.
19. Toni Schmader and Michael Johns, "Converging Evidence that Stereotype Threat Reduces Working Memory Capacity," *Journal of Personality and Social Psychology* 85.3 (2003): 440–452.
20. Diane M. Quinn and Steven J. Spencer, "The Interference of Stereotype Threat with Women's Generation of Mathematical Problem-Solving Strategies," *Journal of Social Issues* 57.1 (2001): 55–71; Steele and Aronson, "Stereotype Threat and the Intellectual Test Performance of African Americans"; Joshua Aronson, Carrie B. Fried, and Catherine Good, "Reducing the Effects of Stereotype Threat on African American College Students by Shaping Theories of Intelligence," *Journal of Experimental Social Psychology* 38.2 (2002): 113–125; Dustin B. Thoman, Paul H. White, Niwako Yamawaki, and Hirofumi Koishi, "Variations of Gender-Math Stereotype Content Affect Women's Vulnerability to Stereotype Threat," *Sex Roles* 58.9–10 (2008): 702–712.
21. David M. Marx and Phillip Atiba Goff, "Clearing the Air: The Effect of Experimenter Race on Target's Test Performance and Subjective Experience," *British Journal of Social Psychology* 44.4 (2005): 645–657; David M. Marx and Jasmin S. Roman, "Female Role Models: Protecting Women's Math Test Performance," *Personality and Social Psychology Bulletin* 28.9 (2002): 1183–1193; Rusty B. McIntyre, Charles G. Lord, Dana M. Gresky, Laura L. Ten Eyck, G. D. Jay Frye, and Charles F. Bond, Jr., "A Social Impact Trend in the Effects of Role Models on Alleviating Women's Mathematics Stereotype Threat," *Current Research in Social Psychology* 10.9 (2005): 116–136; Rusty B. McIntyre, René M. Paulson, and Charles G. Lord, "Alleviating Women's Mathematics Stereotype Threat through Salience of Group Achievements," *Journal of Experimental Social Psychology* 39.1 (2003): 83–90; Geoffrey L. Cohen, Claude M. Steele, and Lee D. Ross, "The Mentor's Dilemma: Providing Critical Feedback across the Racial Divide," *Personality and Social Psychology Bulletin* 25.10 (1999): 1302–1318.
22. Joshua Aronson, Michael J. Lustina, Catherine Good, Kelli Keough, Claude M. Steele, and Joseph Brown, "When White Men Can't Do Math: Necessary and Sufficient Factors in Stereotype Threat," *Journal of Experimental Social Psychology* 35.1 (1999): 29–46; Cynthia M. Frantz, Amy J. C. Cuddy, Molly Burnett, Heidi Ray, and Allen Hart, "A Threat in the Computer: The Race Implicit Association Test as a Stereotype Threat Experience," *Personality and Social Psychology Bulletin* 30.12 (2004): 1611–1124; Jeff Stone, "Battling Doubt by Avoiding Practice: The Effects of Stereotype Threat on Self-Handicapping in White Athletes," *Personality and Social Psychology Bulletin* 28.12 (2002): 1667–1678.

23. Amy K. Kiefer and Denise Sekaquaptewa, "Implicit Stereotypes and Women's Math Performance: How Implicit Gender-Math Stereotypes Influence Women's Susceptibility to Stereotype Threat," *Journal of Experimental Social Psychology* 43.5 (2007a): 825–832; Kiefer and Sekaquaptewa, "Implicit Stereotypes, Gender Identification, and Math-Related Outcomes: A Prospective Study of Female College Students," *Psychological Science* 18.1 (2007b): 13–18.
24. Steele, "A Threat in the Air."
25. Joshua Aronson et al. "When White Men Can't Do Math"; Johannes Keller, "Stereotype Threat in Classroom Settings: The Interactive Effect of Domain Identification, Task Difficulty and Stereotype Threat on Female Students' Math Performance," *British Journal of Educational Psychology* 77.2 (2007): 323–338; Steven J. Spencer, Claude M. Steele, and Diane M. Quinn, "Stereotype Threat and Women's Math Performance," *Journal of Experimental Social Psychology* 35.1 (1999): 4–28.
26. Paul G. Davies, Steven J. Spencer, Diane M. Quinn, and Rebecca Gerhardstein, "Consuming Images: How Television Commercials that Elicit Stereotype Threat Can Restrain Women Academically and Professionally," *Personality and Social Psychology Bulletin* 28.12 (2002): 1615–1628; Osborne and Walker, "Stereotype Threat"; Toni Schmader, Brenda Major, and Richard W. Gramzow, "Coping with Ethnic Stereotypes in the Academic Domain: Perceived Injustice and Psychological Disengagement," *Journal of Social Issues* 57.1 (2001): 93–111.
27. Charles G. Lord and Delia S. Saenz, "Memory Deficits and Memory Surfeits: Differential Cognitive Consequences of Tokenism for Tokens and Observers," *Journal of Personality and Social Psychology* 49.4 (1985): 918–926; Mischa Thompson and Denise Sekaquaptewa, "When Being Different Is Detrimental: Solo Status and the Performance of Women and Racial Minorities," *Analyses of Social Issues & Public Policy* 2.1 (2002): 183–203.
28. Rosabeth Moss Kanter, *Men and Women of the Corporation* (New York: Basic Books, 1977); Yolanda Flores Niemann and John F. Dovidio, "Relationship of Solo Status, Academic Rank, and Perceived Distinctiveness to Job Satisfaction of Racial/Ethnic Minorities," *Journal of Applied Psychology* 83.1 (1998): 55–71; Denise Sekaquaptewa, "On Being the Solo Faculty Member of Color: Research Evidence from Field and Laboratory Studies," in *Engaging Our Faculties*, ed. Stephanie A Fryberg and Ernesto Javier Martínez (manuscript in progress).
29. John F. Dovidio, Samuel L. Gaertner, Yolanda F. Niemann, and Kevin Snider, "Racial, Ethnic, and Cultural Differences in Responding to Distinctiveness and Discrimination on Campus: Stigma and Common Group Identity," *Journal of Social Issues* 57.1 (2001): 167–188.
30. Phillip A. Goff, "Whiteness: How Junior Faculty Negotiate the Unwritten Racial Requests of the Academy," in *Engaging Our Faculties*, ed. Stephanie A. Fryberg and Ernesto Javier Martínez (manuscript in progress).

31. Dovidio, "Racial, Ethnic, and Cultural Differences in Responding to Distinctiveness and Discrimination on Campus"; John F. Dovidio and Samuel L. Gaertner, *Prejudice, Discrimination, and Racism* (Orlando, FL: Academic Press, 1986).
32. Kanter, *Men and Women of the Corporation*; Janice D. Yoder, "Making Leadership Work More Effectively for Women," *Journal of Social Issues* 57.4 (2001): 815–828; Janice D. Yoder, Thomas L. Schleicher, and Theodore W. McDonald, "Empowering Token Women Leaders: The Importance of Organizationally Legitimated Credibility," *Psychology of Women Quarterly* 22.2 (1998): 209–222.
33. Jennifer Crocker and Kathleen M. McGraw, "What's Good for the Goose Is Not Good for the Gander: Solo Status as an Obstacle to Occupational Achievement for Males and Females," *The American Behavioral Scientist* 27.3 (1984): 357–369; Shelley E. Taylor, Susan T. Fiske, Nancy L. Etcoff, and Audrey J. Ruderman, "Categorical and Contextual Bases of Person Memory and Stereotyping," *Journal of Personality and Social Psychology* 36.7 (1978): 778–793.
34. Kanter, *Men and Women of the Corporation*.
35. Marilynn B. Brewer, "The Psychology of Prejudice: Ingroup Love or Outgroup Hate?" *Journal of Social Issues* 55.3 (1999): 429–444; Stewart Shapiro and Mark T. Spence, "Mind over Matter? The Inability to Counteract Contrast Effects Despite Conscious Effort," *Psychology and Marketing* 22.3 (2005): 225–245; Suzanne Swan and Robert S. Wyer, Jr., "Gender Stereotypes and Social Identity: How Being in the Minority Affects Judgment of Self and Others," *Personality and Social Psychology Bulletin* 23.12 (1997): 1265–1276.
36. Brewer, "The Psychology of Prejudice," 429–444; Taylor, Fiske, Etcoff, and Ruderman, "Categorical and Contextual Bases of Person Memory and Stereotyping"; Michael A. Zárate and Eliot R. Smith, "Person Categorization and Stereotyping," *Social Cognition* 8 (1990): 161–185.
37. Phillip Atiba Goff, Claude M. Steele, and Paul G. Davies, "The Space between Us: Stereotype Threat and Distance in Interracial Contexts," *Journal of Personality and Social Psychology* 94.1 (2008): 91–107.
38. Herminia Ibarra, "Race, Opportunity, and Diversity of Social Circles in Managerial Networks," *The Academy of Management Journal* 38.3 (1995): 673–703.
39. Goff, "Whiteness: How Junior Faculty Negotiate the Unwritten Racial Requests of the Academy."
40. Charles Stangor, Christine Carr, and Lisa Kiang, "Activating Stereotypes Undermines Task Performance Expectations," *Journal of Personality and Social Psychology* 75.5 (1998): 1191–1197; Lord and Saenz, "Memory Deficits and Memory Surfeits"; Delia S. Saenz and Charles G. Lord, "Reversing Roles: A Cognitive Strategy for Undoing Memory Deficits Associated with Token Status," *Journal of Personality and Social Psychology* 56.5 (1989): 698–708; Neimann and Dovidio, "Relationship

of Solo Status, Academic Rank, and Perceived Distinctiveness to Job Satisfaction of Racial/Ethnic Minorities."
41. Paul R. Sackett, Cathy L. DuBois, and Ann W. Noe, "Tokenism in Performance Evaluation: The Effects of Work Group Representation on Male-Female and White-Black Differences in Performance Ratings," *Journal of Applied Psychology* 76.2 (1991): 263–267.
42. Neimann and Dovidio, "Relationship of Solo Status, Academic Rank, and Perceived Distinctiveness to Job Satisfaction of Racial/Ethnic Minorities."
43. Mary C. Murphy, Claude M. Steele, and James J. Gross, "Signaling Threat: How Situational Cues Affect Women in Math, Science, and Engineering Settings," *Psychological Science* 18.10 (2007): 879–885; Denise Sekaquaptewa and Mischa Thompson, "Solo Status, Stereotype Threat, and Performance Expectancies: Their Effects on Women's Performance," *Journal of Experimental Social Psychology* 39.1 (2003): 68–74; Stangor, Carr, and Kiang, "Activating Stereotypes Undermines Task Performance Expectations."
44. Sekaquaptewa and Thompson, "Solo Status, Stereotype Threat, and Performance Expectancies."
45. Neimann and Dovidio, "Relationship of Solo Status, Academic Rank, and Perceived Distinctiveness to Job Satisfaction of Racial/Ethnic Minorities"; Sackett, DuBois, and Noe, "Tokenism in Performance Evaluation."
46. Marx and Goff, "Clearing the Air."
47. Frances J. Milliken and Luis L. Martins, "Searching for Common Threads: Understanding the Multiple Effects of Diversity in Organizational Groups," *The Academy of Management Review* 21.2 (1996): 402–433.
48. Nalini Ambady, Sue K. Paik, Jennifer Steele, Ashli Owen-Smith, and Jason P. Mitchell, "Deflecting Negative Self-Relevant Stereotype Activation: The Effects of Individuation," *Journal of Experimental Social Psychology* 40 (2004): 401–408; Matthew S. McGlone and Joshua Aronson, "Stereotype Threat, Identity Salience, and Spatial Reasoning," *Journal of Applied Developmental Psychology* 27.5 (2006): 486–493; Harriet E. S. Rosenthal, Richard J. Crisp, and Mein-Woei Suen, "Improving Performance Expectancies in Stereotypic Domains: Task Relevance and the Reduction of Stereotype Threat," *European Journal of Social Psychology* 37.3 (2007): 586–597.
49. Yoder, "Making Leadership Work More Effectively for Women"; Yoder, "Rethinking Tokenism: Looking Beyond Numbers," *Gender and Society* 5.2 (1991): 178–192; Yoder, Schleicher, and McDonald, "Empowering Token Women Leaders."
50. Hazel Rose Markus, "Pride, Prejudice, and Ambivalence: Toward a Unified Theory of Race and Ethnicity," *American Psychologist* 63.8 (2008): 651–670; Victoria C. Plaut and Hazel Rose Markus, "The

'Inside' Story: A Cultural-Historical Analysis of Being Smart and Motivated, American Style," in *Handbook of Competence and Motivation*, ed. Andrew J. Elliot and Carol S. Dweck (New York: Guilford Press, 2007), pp. 457–488.
51. John Locke, *Essay Concerning Human Understanding* (Oxford: Clarendon Press; New York: Oxford University Press, 1979); Liane M. Davey, D. Ramona Bobocel, Leanne S. Son Hing, and Mark P. Zanna, "Preference for the Merit Principle Scale: An Individual Difference Measure of Distributive Justice Preferences," *Social Justice Research* 12.3 (1999): 223–240.
52. P. H. Nidditch, *John Locke: An Essay Concerning Human Understanding* (Oxford: Clarendon Press, 1975). Jennifer L. Hochschild, "'Succeeding More' and 'under the Spell': Affluent and Poor Blacks' Beliefs about the American Dream," in *Facing Up to the American Dream: Race, Class, and the Soul of the Nation* (Princeton, NJ: Princeton University Press, 1995), pp. 72–88; George Dearborn Spindler and Louise S. Spindler, *The American Cultural Dialogue and Its Transmission* (London; New York: Falmer Press, 1990).
53. Hochschild, "'Succeeding More' and 'under the Spell'"; Stephen J. McNamee and Robert K. Miller, *The Meritocracy Myth* (Lanham, MD: Rowman & Littlefield, 2004); Spindler and Spindler, *The American Cultural Dialogue and Its Transmission*.
54. Hochschild, "'Succeeding More' and 'under the Spell'"; McNamee and Miller, *The Meritocracy Myth*; Davey, Bobocel, Hing, Zanna, "Preference for the Merit Principle Scale."
55. Hazel Rose Markus, Claude M. Steele, and Dorothy M. Steele, "Colorblindness as a Barrier to Inclusion: Assimilation and Nonimmigrant Minorities," *Deadalus* 129 (2000): 233–259.
56. Benjamin Baez, "Race-Related Service and Faculty of Color: Conceptualizing Critical Agency in Academe," *Higher Education* 39.3 (2000): 363–391.
57. William E. B. DuBois and Carter Burden, *The Souls of Black Folk* (Chicago, IL: A.C. McClurg & Co., 1903).
58. Robyn K. Mallet and Janet K. Swim, "The Influence of Inequality, Responsibility and Justifiability on Reports of Group-Based Guilt for Ingroup Privilege," *Group Processes & Intergroup Relations* 10.1 (2007): 57–69; Janet K. Swim and Deborah L. Miller, "White Guilt: Its Antecedents and Consequences for Attitudes toward Affirmative Action," *Personality and Social Psychology Bulletin* 25.4 (1999): 500–514.
59. Patricia Gurin, E. Dey, S. Hurtado, and G. Gurin, "Diversity and Higher Education: Theory and Impact on Educational Outcomes," *Harvard Educational Review* 72.3 (2002): 330.
60. Monisha Bajaj, "Lanterns and Street Signs: Effective Mentoring for Greater Equality in the Academy," in *Engaging Our Faculties* ed. Stephanie A. Fryberg and Ernesto Javier Martínez (manuscript in progress).

BIBLIOGRAPHY

ACT Summary Report for Graduates 1998–99 to 2002–03. Detroit Public Schools, 2004.

Adelman, C. "Using the Modular System of Record-Keeping to Track Changes in Delivered Knowledge." September 21, 2001. Paper presented at the 2001 Forum of the European Association for Institutional Research, Porto, Portugal (September 21, 2001).

Adelman, Clifford, and United States. Office of Educational Research and Improvement. *The New College Course Map and Transcript Files: Changes in Course-Taking and Achievement, 1972–1993: Based on the Postsecondary Records from Two National Longitudinal Studies.* Washington, DC: U.S. Dept. of Education, Office of Educational Research and Improvement; For sale by the U.S. G.P.O., Supt. of Docs., 1999.

"Admissions Research: News and Information." *The University of Texas at Austin.* (2007). http://www.utexas.edu/student/admissions/research/.

Agathangelou, Anna M., and Lily H. M. Ling. "An Unten(Ur)Able Position: The Politics of Teaching for Women of Color in the US." *International Feminist Journal of Politics* 4.3 (2002): 368–398.

An Agenda for Excellence: Creating Flexibility in Tenure-Track Faculty Careers. American Council on Education, Office of Women in Higher Education, February 2005.

Allen, Danielle. *Talking to Strangers: Anxieties of Citizenship since Brown v Board of Education.* Chicago, IL: University of Chicago Press, 2006.

Altbach, Philip G., Robert Oliver Berdahl, and Patricia J. Gumport. *American Higher Education in the Twenty-First Century: Social, Political, and Economic Challenges.* Baltimore, MD: Johns Hopkins University Press, 1999.

Ambady, Nalini, Sue K. Paik, Jennifer Steele, Ashli Owen-Smith, and Jason P. Mitchell. "Deflecting Negative Self-Relevant Stereotype Activation: The Effects of Individuation." *Journal of Experimental Social Psychology* 40 (2004): 401–408.

"America's Promise Alliance Launches National Campaign to Combat Nation's High School Dropout and College-Readiness Crisis." April 1, 2008. http://www.americaspromise.org/About-the-Alliance/Press-Room/Press-Releases/2008/2008-April-01-Americas-Promise-Alliance-Launches-National-Campaign.aspx (accessed February 5, 2010).

America's Untapped Resource: Low-Income Students in Higher Education. Ed. Richard D. Century Foundation Kahlenberg. New York: Century Foundation Press, 2004.

American Council on Education. *Minorities in Higher Education 2008: 23rd Status Report* (2008).

"Amherst College Will Replace Loans with Scholarships in Financial Aid Packages for All Students Beginning in 2008–09." *Amherst College* (July 19, 2007). Accessed February 13, 2008.

Ancheta, Angelo N. "Revisiting Bakke and Diversity-Based Admissions: Constitutional Law, Social Science Research, and the University of Michigan Affirmative Action Cases. Briefing Paper." (2003).

Anderson, G. M. *Building a People's University in South Africa: Race, Compensatory Education, and the Limits of Democratic Reform.* New York: P. Lang, 2002.

———. "In the Name of Diversity: Education and the Commoditization and Consumption of Race in the United States." *Urban Review New York* 37.5 (2005): 399–423.

Anderson, Gregory, M. A., Eleanor J. B. Daugherty, and Darlene M. Corrigan. "The Search for a Critical Mass of Minority Students: Affirmative Action and Diversity at Highly Selective Universities and Colleges." *PEGS: Committee on the Political Economy of the Good Society* 14.3 (Spring 2006): 51–57.

Andrews, Nancy C. "Climbing through Medicine's Glass Ceiling." *The New England Journal of Medicine* 357.19 (November 2007): 1887–1889.

Angelique, Holly, Ken Kyle, and Ed Taylor. "Mentors and Muses: New Strategies for Academic Success." *Innovative Higher Education* 26.3 (2002): 195–209.

Antonio, Anthony Lising. "Faculty of Color Reconsidered: Reassessing Contributions to Scholarship." *The Journal of Higher Education* 73.5 (2002): 582–602.

Arenson, Karen W. "Study of College Readiness Finds No Progress in Decade." *New York Times.* October 14, 2004: A. 26.

Aronson, Joshua, Carrie B. Fried, and Catherine Good. "Reducing the Effects of Stereotype Threat on African American College Students by Shaping Theories of Intelligence." *Journal of Experimental Social Psychology* 38.2 (2002): 113–125.

Aronson, Joshua, Michael J. Lustina, Catherine Good, Kelli Keough, Claude M. Steele, and Joseph Brown. "When White Men Can't Do Math: Necessary and Sufficient Factors in Stereotype Threat." *Journal of Experimental Social Psychology* 35.1 (1999): 29–46.

"Arthur O. Eve Higher Education Opportunity Program (HEOP)." Office of Higher Education, New York State Education Department. http://www.highered.nysed.gov/kiap/colldev/HEOP/ (accessed February 5, 2010).

Atkinson, Richard. "Diversity: Not There Yet." *Washington Post.* April 20, 2003.

BIBLIOGRAPHY

Attewell, Paul A. *Passing the Torch: Does Higher Education for the Disadvantaged Pay Off across the Generations?* Ed. David E. Lavin. New York: Russell Sage Foundation, 2007.

Baez, Benjamin. "Race-Related Service and Faculty of Color: Conceptualizing Critical Agency in Academe." *Higher Education* 39.3 (2000): 363–391.

Bajaj, Monisha. "Lanterns and Street Signs: Effective Mentoring for Greater Equality in the Academy." *Engaging Our Faculties: Junior Faculty of Color and Senior University Administrators on Diversity and Excellence in Higher Education.* Ed. Stephanie A. Fryberg and Ernesto Javier Martínez (manuscript in progress).

Bane, Mary Jo, and David T. Ellwood. "Slipping Into and Out of Poverty: The Dynamics of Spells." *The Journal of Human Resources* 21.1 (1986): 1.

Barbara Grutter v. Lee Bollinger et al. 539 U.S. 306. Supreme Court of the United States. 2003.

Baumeister, Roy F., and Mark R. Leary. "The Need to Belong: Desire for Interpersonal Attachments as a Fundamental Human Motivation." *Psychological Bulletin* 117.3 (1995): 497.

Bell, Derrick A. *Silent Covenants: Brown v. Board of Education and the Unfulfilled Hopes for Racial Reform.* Oxford; New York: Oxford University Press, 2004.

Berlin, Isaiah. "The Crooked Timber of Humanity." *Chapters in the History of Ideas.* Ed. Henry Hardy. New York: Vintage Books, 1992.

Best Colleges Index: National Universities. U.S. News and World Report, September 11, 2001.

"Black Students Less Likely to Take A.P. Exams." *The New York Times.* February 4, 2009: A 19.

Bok, Derek. "Closing the Nagging Gap in Minority Achievement." *The Chronicle of Higher Education* 50.9 (2003): B.20.

Bowen, William G. "Race-Sensitive Admissions: Back to Basics." *The 2002 Andrew W. Mellon Foundation Annual Report.* http://www.mellon.org/news_publications/annual-reports-essays/annual-reports/content2002.pdf (accessed October 9, 2008).

Bowen, William G., and Derek Curtis Bok. *The Shape of the River: Long-Term Consequences of Considering Race in College and University Admissions.* Princeton, NJ: Princeton University Press, [2000], c1998.

Bowen, William G., Martin Kurzweil, and Eugene Tobin. *Equity and Excellence in American Higher Education.* Charlottesville, VA: University of Virginia Press, 2005.

Bowen, William G., Matthew M. Chingos, and Michael S. McPherson. "Helping Students Finish the 4-Year Run." *The Chronicle of Higher Education* (September 8, 2009). http://chronicle.com/article/Helping-Students-Finish-the/48329/.

Bowles, Samuel, and Herbert Gintis. *Schooling in Capitalist America: Educational Reform and the Contradictions of Economic Life.* New York: Basic Books, 1976.

Boyle, Peg, and Bob Boice. "Systematic Mentoring for New Faculty Teachers and Graduate Teaching Assistants." *Innovative Higher Education* 22.3 (1998): 157–179.
Brewer, Marilynn B. "The Psychology of Prejudice: Ingroup Love or Outgroup Hate?" *Journal of Social Issues* 55.3 (1999): 429–444.
———. "The Social Self: On Being the Same and Different at the Same Time." *Personality and Social Psychology Bulletin* 17.5 (1991): 475–482.
Brickman, P., and D. Coates. "Commitment and Mental Health." *Commitment, Conflict, and Caring*. Ed. P. Brickman. Englewood Cliffs, NJ: Prentice-Hall, 1987, 222–309.
Brief as Amici Curiae Supporting Respondents, Gratz v. Bollinger. 2003.
Brief as Amici Curiae Supporting Respondents, Grutter v. Bollinger. 2003.
Brooks, David. "The Insider's Crusade." *New York Times*. November 21, 2008: A.35.
Brown University. *Response of Brown University to the Report of the Steering Committee on Slavery and Justice*. Providence, RI: Brown University, 2007.
Cantor, Nancy. "Re: The Evidence of Things Not Seen" (*Wall Street Journal*, May 16, 2003) by Chetly Zarko." *Chicago Sun Times*. June 22, 2003.
———. "Valuing Public Scholarship." *The Presidency: The American Council on Education Magazine for Higher Education Leaders* (Spring 2005): 35, 35–37.
Cantor, Nancy, and Steven D. Lavine. "Taking Public Scholarship Seriously." *The Chronicle of Higher Education* 52.40 (2006): B.20.
Cantor, Nancy, and Catherine Sanderson. "Life Task Participation and Well-Being: The Importance of Taking Part in Daily Life." *Well-Being: The Foundations of Hedonic Psychology*. Ed. D. Kahneman, E. Diener, and N. Schwarz. New York: The Russell Sage Foundation, 1999, 230–243.
Cantor, Nancy, and Steven Schomberg. "Poised between 2 Worlds: The University as Monastery and Marketplace." *Educause Review* 38.2 (2003): 12.
Cantor, Nancy, M. Kemmelmeier, J. Basten, and D. Prentice. "Life Task Pursuit in Social Groups: Balancing Self-Exploration and Social Integration." *Self and Identity* 1.2 (April 2002): 177, 177–184.
Cantor, Nancy, and University of Illinois at Urbana-Champaign. Institute of Government and Public Affairs. *Higher Education Policy-Making in the Melting Pot of Stakeholder Voices: The Michigan Affirmative Action Cases*. Urbana, IL: Institute of Government and Public Affairs, University of Illinois, 2004.
Castelvecchi, David. "Numbers Don't Add Up for U.S. Girls." *Science News* 174.10 (2008): 10.
Center for Educational Performance and Information. *2003–2004 Michigan Graduation/Dropout Rates*. State of Michigan, 2005.
Chang, Mitchell J. "The Diversity Vote: The Michigan Civil Rights Initiative." *OpEdNews*. October, 17, 2006.

Chang, Mitchell J, Daria Witt, James Jones, and Kenji Hakuta. *Compelling Interest: Examining the Evidence on Racial Dynamics in Colleges and Universities*. Stanford, CA: Stanford Education, 2003.

Clive, Cookson. "Poverty Mars Formation of Infant Brains." *Financial Times*. February 16, 2008: 5.

Cohen, Geoffrey L., Claude M. Steele, and Lee D. Ross. "The Mentor's Dilemma: Providing Critical Feedback across the Racial Divide." *Personality and Social Psychology Bulletin* 25.10 (1999): 1302–1318.

Cokley, Kevin O. "Ethnicity, Gender and Academic Self-Concept: A Preliminary Examination of Academic Disidentification and Implications for Psychologists." *Cultural Diversity and Ethnic Minority Psychology* 8.4 (2002): 378–388.

Comfort v. Lynn School Committee. 126 U.S. 798. Supreme Court of the United States. 2005.

Crocker, Jennifer, and Kathleen M. McGraw. "What's Good for the Goose Is Not Good for the Gander: Solo Status as an Obstacle to Occupational Achievement for Males and Females." *The American Behavioral Scientist* 27.3 (1984): 357–369.

Crocker, Jennifer, Brenda Major, and Claude M. Steele. "Social Stigma." *The Handbook of Social Psychology*. Ed. Daniel Todd Gilbert, Susan T. Fiske, and Gardner Lindzey. 4th ed. Boston; New York: McGraw-Hill; Distributed exclusively by Oxford University Press, 1998, 504–553.

D'Andrade, Roy Goodwin. "The Cultural Part of Cognition." *Cognitive Science* 5.3 (1981): 179–195.

———. "Moral Models in Anthropology." *Current Anthropology* 36.3 (1995): 399–408.

Davey, Liane M. D. Ramona Bobocel, Leanne S. Son Hing, and Mark P. Zanna. "Preference for the Merit Principle Scale: An Individual Difference Measure of Distributive Justice Preferences." *Social Justice Research* 12.3 (1999): 223–240.

Davies, Paul G. Steven J. Spencer, Diane M. Quinn, and Rebecca Gerhardstein. "Consuming Images: How Television Commercials that Elicit Stereotype Threat Can Restrain Women Academically and Professionally." *Personality and Social Psychology Bulletin* 28.12 (2002): 1615–1628.

Defending Diversity: Affirmative Action at the University. Ed. Patricia Lehman Gurin Jeffrey S. Ann Arbor, MI: University of Michigan Press, 2004.

"Descriptor plus." *College Board* (2007). http://professionals.collegeboard.com/higher-ed/recruitment/descriptor-plus.

Dougherty, Kevin J. *The Contradictory College: The Conflicting Origins, Impacts, and Futures of the Community College*. Albany, NY: State University of New York Press, 1994.

Dovidio, John F., and Samuel L. Gaertner. *Prejudice, Discrimination, and Racism*. Orlando, FL: Academic Press, 1986.

Dovidio, John F., Samuel L. Gaertner, Yolanda F. Niemann, and Kevin Snider. "Racial, Ethnic, and Cultural Differences in Responding to

Distinctiveness and Discrimination on Campus: Stigma and Common Group Identity." *Journal of Social Issues* 57.1 (2001): 167–188.

DuBois, William E. B., and Carter Burden. *The Souls of Black Folk*. Chicago, IL: A.C. McClurg & Co., 1903.

Editorial Projects in Education Research Center. *Cities in Crisis: A Special Analytic Report on High School Graduation*. America's Promise Alliance, April 1, 2008.

Ellison, Julie. "Director's Column." *Imagining America Newsletter*. Summer 2004.

Equity and Excellence in American Higher Education. Brookings Institution, Miller Reporting Co., Inc., 2005.

Fabrikant, Geraldine. "Colleges Struggle to Preserve Financial Aid." *New York Times*. November 11, 2008: F.25.

"The Fifth Circuit Decision on Affirmative Action Appealed to the Supreme Court." *Civil Rights Monitor* 8.4 (Winter 1996). http://www.civilrights.org/monitor/vol8_no4/art2.html (accessed February 5, 2010).

"Financial Aid." *Williams College*. February 21, 2008. http://www.williams.edu/admission/finaid.php.

Fiske, Alan Page, Shinobu Kitayama, Hazel Rose Markus, and Richard Nisbett. "The Cultural Matrix of Social Psychology." *The Handbook of Social Psychology*. Ed. Daniel Todd Gilbert, Susan T. Fiske, and Gardner Lindzey. 4th ed. Boston; New York: McGraw-Hill; Distributed exclusively by Oxford University Press, 1998, 915–981.

Fox, Mary Jo Tippeconnic. "American Indian Women in Academia: The Joys and Challenges." *Journal about Women in Higher Education* 1 (2008): 202–221.

Frank, David John. "Rethinking History: Change in the University Curriculum: 1910–90." *Sociology of Education* 67.4 (1994): 231–242.

Frank, David John, Suk-Ying Wong, John Meyer, and Francisco Ramirez. "What Counts as History: A Cross-National and Longitudinal Study of University Curricula." *Comparative Education Review* 44.1 (2000): 29.

Frankenberg, Erica, Chungmei Lee, and Gary Orfield. "A Multiracial Society with Segregated Schools: Are We Losing the Dream?" (January 2003). http://www.civilrightsproject.ucla.edu/research/reseg03/reseg03_full.php; http://harvardscience.harvard.edu.proxy.library.cornell.edu/culture-society/articles/multiracial-society-segregated-schools.

Frantz, Cynthia M., Amy J. C. Cuddy, Molly Burnett, Heidi Ray, and Allen Hart. "A Threat in the Computer: The Race Implicit Association Test as a Stereotype Threat Experience." *Personality and Social Psychology Bulletin* 30.12 (2004): 1611–1124.

Freeman, Kassie. *African American Culture and Heritage in Higher Education Research and Practice*. Westport, CT: Praeger, 1998.

Frumkin, Howard, Lawrence Frank, and Richard Joseph Jackson. *Urban Sprawl and Public Health: Designing, Planning, and Building for Healthy Communities*. Washington, DC: Island Press, 2004.

Howard, John R. "Affirmative Action in Historical Perspective." *Affirmative Action's Testament of Hope: Strategies for a New Era in Higher Education.* Ed. Mildred Garcia. Albany, NY: State University of New York Press, 1997.

Fryberg, Stephanie A., and Hazel Rose Markus. "Cultural Models of Education in American Indian, Asian American and European American Contexts." *Social Psychology of Education* 10.2 (2007): 213–246.

Gage, Matilda Joslyn. *Woman, Church and State: A Historical Account of the Reminiscences of Matriarchate.* Ed. Sally Roesch Wagner. Aberdeen, SD: Sky Carrier Press, 1998 [1893].

García, Mildred. *Affirmative Action's Testament of Hope: Strategies for a New Era in Higher Education.* Albany, NY: State University of New York Press, 1997.

Glater, Jonathan D. "House Passes Bill Aimed at College Costs." *New York Times.* February 8, 2008: A.13.

Goff, Phillip A. "Whiteness: How Junior Faculty Negotiate the Unwritten Racial Requests of the Academy." *Engaging Our Faculties: Junior Faculty of Color and Senior University Administrators on Diversity and Excellence in Higher Education.* Ed. Stephanie A. Fryberg and Ernesto Javier Martínez (manuscript in progress).

Goff, Phillip Atiba, Claude M. Steele, and Paul G. Davies. "The Space between Us: Stereotype Threat and Distance in Interracial Contexts." *Journal of Personality and Social Psychology* 94.1 (2008): 91–107.

Golden, Daniel. *The Price of Admission: How America's Ruling Class Buys Its Way into Elite Colleges—and Who Gets Left Outside the Gates.* 1st ed. New York: Crown Publishers, 2006.

Gratz v. Bollinger. 539 U.S. 244. Supreme Court of the United States. 2003.

Greene, Howard, and Matthew Greene. "Need-Blind Admissions Policies: What They Mean to Your Pocketbook." *Peterson's.* http://www.petersons.com/common/article.asp?id=3269&path=ug.pfs.advice&sponsor=1 (accessed February 5, 2010).

Grutter v. Bollinger. 539 U.S. 306. Supreme Court of the United States. 2003.

Gurin, Patricia. *Educational Benefits of Intergroup Dialogue.* Unpublished grant proposal. Ann Arbor, MI: University of Michigan, August 2004.

Gurin, Patricia, E. Dey, S. Hurtado, and G. Gurin. "Diversity and Higher Education: Theory and Impact on Educational Outcomes." *Harvard Educational Review* 72.3 (2002): 330.

Gurin, Patricia, Biren (Ratnesh) A. Nagda, and Gretchen E. Lopez. "The Benefits of Diversity in Education for Democratic Citizenship." *The Journal of Social Issues* 60.1 (2004): 17.

Gurin, Patricia, University of Michigan, and Cooperative Institutional Research Program (U.S.). *Expert Report of Patricia Gurin: Gratz Et Al v. Bollinger, Et Al., no. 97–75321 (E.D. Mich.); Grutter, Et Al. v. Bollinger,*

Et Al., no. 97–75928 (E.D. Mich.). Ann Arbor, MI: University of Michigan, 1999.

Hannum, Emily, Jere Behrman, and Meiyan Wang. *Human Capital in China*. Paper prepared for the conference "China's Economic Transition: Origins, Outcomes, Mechanisms and Consequences." Pittsburgh, PA, November 4–7, 2004.

"Harvard Announces Sweeping Middle-Income Initiative." *Harvard University Gazette*. December 10, 2007.

Harvard College Admissions Office. "Harvard Student Recruitment Program." http://www.admissions.college.harvard.edu.proxy.library.cornell.edu/prospective/hrp/index.html.

"The Haudenosaunee Promise at Syracuse University." *Syracuse University*. http://financialaid.syr.edu/scholar-haudenosauneeflyer.htm.

Healy, P. "Beneath Campaign Surface, Obama's Race Remains a Potent Issue." *The International Herald Tribune*. October 13, 2008.

Heckman, James J. "Skill Formation and the Economics of Investing in Disadvantaged Children." *Science* 312.5782 (June 2006): 1900–1902.

Hentoff, Nat. "Affirmative Action Discords." *Jewish World Review* May 28, 2003. http://www.jewishworldreview.com/cols/hentoff052803.asp (accessed February 5, 2010).

Hill, Catharine, Gordon Winston, and Stephanie Boyd. *Affordability: Family Incomes and Net Prices at Highly Selective Private Colleges and Universities*. Discussion Paper no. 66. Williams College Project on the Economics of Higher Education, October 2003.

Hirsch, Arnold R. *Making the Second Ghetto: Race and Housing in Chicago, 1940–1960*. Cambridge [Cambridgeshire]; New York: Cambridge University Press, 1983.

Hi-Voltage Wire Works, Inc. v. City of San Jose. 2000.

Hochschild, Jennifer L. "'Succeeding More' and 'under the Spell': Affluent and Poor Blacks' Beliefs about the American Dream." *Facing Up to the American Dream: Race, Class, and the Soul of the Nation*. Princeton, NJ: Princeton University Press, 1995, 72–88.

Holland, Dorothy C., and Naomi Quinn. *Cultural Models in Language and Thought*. Cambridge [Cambridgeshire]; New York: Cambridge University Press, 1987.

Horowitz, David. "In Defense of Intellectual Diversity." *The Chronicle of Higher Education* 50.23 (2004): B.12.

Hosseini, Khaled. *The Kite Runner*. New York: Riverhead Books, 2004.

Hurtado, Sylvia. "The Campus Racial Climate: Contexts of Conflict." *The Journal of Higher Education* 63.5 (1992): 539–569.

———. "Linking Diversity with the Educational and Civic Missions of Higher Education." *The Review of Higher Education* 30.2 (2006): 185–196.

Hurtado, Sylvia, et al. *Enacting Diverse Learning Environments: Improving the Climate for Racial/Ethnic Diversity in Higher Education. ASHE-ERIC Higher Education Report, Vol. 26, no. 8*. ERIC Clearinghouse on

Higher Education, Washington, DC. Hurtado, Sylvia, Jeffrey F. Milem, Alma R. Clayton-Pedersen, and Walter R. Allen. "Enhancing Campus Climates for Racial/Ethnic Diversity: Educational Policy and Practice." *The Review of Higher Education* 21.3 (1998): 279–302.

Ibarra, Herminia. "Race, Opportunity, and Diversity of Social Circles in Managerial Networks." *The Academy of Management Journal* 38.3 (1995): 673–703.

Iceland, John, and Rima Wilkes. "Does Socioeconomic Status Matter? Race, Class, and Residential Segregation." *Social Problems* 53.2 (2006): 248.

Intergroup Dialogue: Deliberative Democracy in School, College, Community, and Workplace. Ed. David Louis Schoem and Sylvia Hurtado. Ann Arbor, MI: University of Michigan Press, 2001.

International Challenges to American Colleges and Universities: Looking Ahead. American Council on Education Series on Higher Education. Ed. Katherine Hanson and Joel W. Meyerson, 1995.

Investing in People: Developing All of America's Talent on Campus and in the Workplace. Washington, DC: Business-Higher Education Forum, 2002.

Jackson, D. "McCain Sides with Ban on Affirmative Action." *USA Today*. July 27, 2008.

Jencks, Christopher, and Meredith Phillips. *The Black-White Test Score Gap.* Washington, DC: Brookings Institution Press, 1998.

Jennifer Gratz and Patrick Hamacher v. Lee Bollinger Et Al. 539 U.S. 244. 2003.

Kahneman, Daniel, Ed Diener, and Norbert Schwarz. *Well-Being: The Foundations of Hedonic Psychology.* New York: Russell Sage Foundation, 1999.

Kane, Thomas J. *The Price of Admission: Rethinking How Americans Pay for College.* Washington, DC: Brookings Institution Press, 1999.

Kanter, Rosabeth Moss. *Men and Women of the Corporation.* New York: Basic Books, 1977.

Karabel, Jerome. *The Chosen: The Hidden History of Admission and Exclusion at Harvard, Yale, and Princeton.* Boston, MA: Houghton Mifflin Co., 2005.

———. "Community Colleges and Social Stratification." *Harvard Educational Review* 42 (1972): 521–562.

———. "Protecting the Portals: Class and the Community College." *Social Policy* 5.1 (1974): 12–19.

Keller, Johannes. "Stereotype Threat in Classroom Settings: The Interactive Effect of Domain Identification, Task Difficulty and Stereotype Threat on Female Students' Math Performance." *British Journal of Educational Psychology* 77.2 (2007): 323–338.

Kiefer, Amy K., and Denise Sekaquaptewa. "Implicit Stereotypes and Women's Math Performance: How Implicit Gender-Math Stereotypes Influence Women's Susceptibility to Stereotype Threat." *Journal of Experimental Social Psychology* 43.5 (2007a): 825–832.

Kiefer, Amy K., and Denise Sekaquaptewa. "Implicit Stereotypes, Gender Identification, and Math-Related Outcomes: A Prospective Study of Female College Students." *Psychological Science* 18.1 (2007b): 13–18.

Kolata, Gina. "A Surprising Secret to a Long Life: Stay in School." *New York Times*. January 3, 2007: A.1.

Kozulin, Alex. *Psychological Tools: A Sociocultural Approach to Education*. Cambridge: Harvard University Press, 1998.

Kristof, Nicholas. "Wretched of the Earth." *The New York Review of Books* 54 (May 31, 2007).

Krueger, Alan, Jesse Rothstein, and Sarah Turner. *Race, Income and College in 25 Years: The Continuing Legacy of Segregation and Discrimination*. Working Paper #9. Education Research Section, Princeton University, December 2004.

Lamont, Michelle, and Annette Lareau. "Cultural Capital: Allusions, Gaps, and Glissandos in Recent Theoretical Developments." *Sociological Theory* 6 (1988): 153–168.

Lareau, Annette. *Home Advantage: Social Class and Parental Intervention in Elementary Education*. Lanham, MD: Rowman & Littlefield Publishers, 2000.

———. "Social Class Differences in Family-School Relationships: The Importance of Cultural Capital." *Sociology of Education* 60.2 (1987): 73–85.

Lareau, Annette, and Erin McNamara Horvat. "Moments of Social Inclusion and Exclusion: Race, Class, and Cultural Capital in Family-School Relationships." *Sociology of Education* 72.1 (1999): 37.

Lavin, David E., and David Hyllegard. *Changing the Odds: Open Admissions and the Life Chances of the Disadvantaged*. New Haven, CT: Yale University Press, 1996.

Laycock, Douglas. "The Broader Case for Affirmative Action: Desegregation, Academic Excellence, and Future Leadership." *Tulane Law Review* 78.6 (June 2004): 1767–1842.

Lee, Yueh-Ting, and Victor Ottati. "Perceived In-Group Homogeneity as a Function of Group Membership Salience and Stereotype Threat." *Personality and Social Psychology Bulletin* 21.6 (1995): 610–619.

Leonhardt, David. "The New Affirmative Action." *The New York Times*. September 30, 2007.

Levin, J. S. "Global Culture and the Community College." *Community College Journal of Research and Practice* 26.2 (2002): 121–145.

———. *Globalizing the Community College: Strategies for Change in the Twenty-First Century*. New York: Palgrave, 2001.

Levin, Sarah A., and James L. Shulman. *Reclaiming the Game: College Sports and Educational Values*. Princeton, NJ: Princeton University Press, 2003.

Levine, Arthur, and Jana Nidiffer. *Beating the Odds: How the Poor Get to College*. San Francisco, CA: Jossey-Bass Publishers, 1996.

Levine, Arthur and Jeanette Cureton. *When Hope and Fear Collide: A Portrait of Today's College Students.* San Francisco, CA: Jossey-Bass Publishers, 1998.

Lewin, T. "College May Become Unaffordable for Most Americans, Report Says." *The New York Times.* December 3, 2008.

Li, Jin. "U.S and Chinese Cultural Beliefs about Learning." *Journal of Educational Psychology* 95.2 (2003): 258–267.

Lloyd, Kim, Kevin Leicht, and Teresa A. Sullivan. "Minority College Aspirations, Expectations, and Applications under the Texas Top 10% Law." *Social Forces* (March 2008): 1105–1138.

Locke, John. *Essay Concerning Human Understanding.* Oxford: Clarendon Press; New York: Oxford University Press, 1979.

Long, M., and M. Tienda. "Winners and Losers: Changes in Texas University Admissions Post-Hopwood." *Education Evaluation Policy Analysis* 30.3 (September 2008): 255–280.

Lord, Charles G., and Delia S. Saenz. "Memory Deficits and Memory Surfeits: Differential Cognitive Consequences of Tokenism for Tokens and Observers." *Journal of Personality and Social Psychology* 49.4 (1985): 918–926.

Loury, Glenn C. *The Anatomy of Racial Inequality.* Cambridge, MA: Harvard University Press, 2002.

Lowe, Eugene Y. *Promise and Dilemma: Perspectives on Racial Diversity and Higher Education.* Princeton, NJ: Princeton University Press, 1999.

Macdonald, Amie A. "Racial Authenticity and White Separatism: The Future of Racial Program Housing on College Campuses." *Reclaiming Identity: Realist Theory and the Predicament of Postmodernism.* Ed. Paula M. L. Moya and Michael Hames-Garcia. Berkeley, CA: University of California Press, 2000.

Major, Brenda, and Toni Schmader. "Coping with Stigma through Psychological Disengagement." *Prejudice : The Target's Perspective.* Ed. Janet K. Swim and Charles Stangor. San Diego, CA: Academic Press, 1998, 219–241.

Major, Brenda, Steven Spencer, Toni Schmader, Connie Wolfe, and Jennifer Crocker. "Coping with Negative Stereotypes about Intellectual Performance: The Role of Psychological Disengagement." *Personality and Social Psychology Bulletin* 24.1 (1998): 34–50.

Mallett, Robyn K., and Janet K. Swim. "The Influence of Inequality, Responsibility and Justifiability on Reports of Group-Based Guilt for Ingroup Privilege." *Group Processes & Intergroup Relations* 10.1 (2007): 57–69.

Mann, Charles C. "The Founding Sachems." *New York Times.* July 4, 2005: A.13.

Markus, Hazel Rose. "Pride, Prejudice, and Ambivalence: Toward a Unified Theory of Race and Ethnicity." *American Psychologist* 63.8 (2008): 651–670.

Markus, Hazel Rose, Claude M. Steele, and Dorothy M. Steele. "Colorblindness as a Barrier to Inclusion: Assimilation and Nonimmigrant Minorities." *Deadalus* 129 (2000): 233–259.

Markus, Hazel Rose, Patricia R. Mullally, and Shinobu Kitayama. "Selfways: Diversity in Modes of Cultural Participation." *The Conceptual Self in Context: Culture, Experience, Self-Understanding*. Ed. Ulric Neisser and David A. Jopling. Cambridge, UK; New York: Cambridge University Press, 1997, 13–61.

Marx, David M., and Phillip Atiba Goff. "Clearing the Air: The Effect of Experimenter Race on Target's Test Performance and Subjective Experience." *British Journal of Social Psychology* 44.4 (2005): 645–657.

Marx, David M., and Jasmin S. Roman. "Female Role Models: Protecting Women's Math Test Performance." *Personality and Social Psychology Bulletin* 28.9 (2002): 1183–1193.

Marx, Karl. *Capital*. Moscow: Progress Publishers, 1986, 43.

———. *Genesis of Capital*. Moscow: Progress Publishers, 1969.

Mason, Patrick I., Samuel L. Myers Jr., and William Darity Jr. "Is There Racism in Economic Research?" *European Journal of Political Economy* 21 (2005): 755–761.

Massey, Douglas S. *Categorically Unequal: The American Stratification System*. New York: Russell Sage Foundation, 2007.

Massey, Douglas S., Camille Z. Charles, Garvey Lundy, and Mary J. Fischer. *The Source of the River: The Social Origins of Freshmen at America's Selective Colleges and Universities*. Princeton, NJ: Princeton University Press, 2003.

Massey, Douglas S., and Nancy A. Denton. *American Apartheid: Segregation and the Making of the Underclass*. Cambridge, MA: Harvard University Press, 1993.

McGlone, Matthew S., and Joshua Aronson. "Stereotype Threat, Identity Salience, and Spatial Reasoning." *Journal of Applied Developmental Psychology* 27.5 (2006): 486–493.

McIntyre, Rusty B., René M. Paulson, and Charles G. Lord. "Alleviating Women's Mathematics Stereotype Threat through Salience of Group Achievements." *Journal of Experimental Social Psychology* 39.1 (2003): 83–90.

McIntyre, Rusty B., Charles G. Lord, Dana M. Gresky, Laura L. Ten Eyck, G. D. Jay Frye, and Charles F. Bond, Jr. "A Social Impact Trend in the Effects of Role Models on Alleviating Women's Mathematics Stereotype Threat." *Current Research in Social Psychology* 10.9 (2005): 116–136.

McKinney, W. L. "When Hope and Fear Collide: A Portrait of Today's College Student." *Choice* 36.2 (1998): 369.

McNamee, Stephen J., and Robert K. Miller. *The Meritocracy Myth*. Lanham, MD: Rowman & Littlefield, 2004.

McPherson, Michael S., and Morton Owen Schapiro, eds. *College Access: Opportunity or Privilege?* New York: The College Board, 2006.

Miller, Sara. "Haudenosaunee Promise Succeeds in Helping Native American Students Attend SU." *Syracuse University News.* August 22, 2006. http://sunews.syr.edu/story_details.cfm?id=3428.

Milliken, Frances J., and Luis L. Martins. "Searching for Common Threads: Understanding the Multiple Effects of Diversity in Organizational Groups." *The Academy of Management Review* 21.2 (1996): 402–433.

"Mini-Digest of Education Statistics 2006." *U.S. Department of Education National Center.* http://nces.ed.gov/pubs2007/2007067.pdf.

Morzinski, Jeffrey A., Sabina Diehr, David J. Bower, and Deborah E. Simpson. "A Descriptive, Cross-Sectional Study of Formal Mentoring for Faculty." *Family Medicine* 28.6 (1996): 434–438.

Mun, Tsang C. "Educational and National Development in China since 1949: Oscillating Policies and Enduring Dilemmas." *China Review* (2000): 579–618.

Murphy, Mary C., Claude M. Steele, and James J. Gross. "Signaling Threat: How Situational Cues Affect Women in Math, Science, and Engineering Settings." *Psychological Science* 18.10 (2007): 879–885.

National Center for Education Statistics, 2001. U.S. Census Bureau, 2002.

National Educational Longitudinal Study (NELS). Education Statistics, 2001.

Nidditch, P. H. *John Locke: An Essay Concerning Human Understanding.* Oxford: Clarendon Press, 1975.

Niemann, Yolanda Flores, and John F. Dovidio. "Relationship of Solo Status, Academic Rank, and Perceived Distinctiveness to Job Satisfaction of Racial/Ethnic Minorities." *Journal of Applied Psychology* 83.1 (1998): 55–71.

Niu, Sunny Xinchun, Teresa Sullivan, and Marta Tienda. "Minority Talent Loss and the Texas Top 10 Percent Law." *Social Science Quarterly* 89.4 (2008): 831.

Nussbaum, A. David, and Claude M. Steele. "Situational Disengagement and Persistence in the Face of Adversity." *Journal of Experimental Social Psychology* 43.1 (2007): 127–134.

Office of Public Affairs. "Yale Cuts Costs for Families and Students." *Yale University.* February 13, 2008. http://www.yale.edu.proxy.library.cornell.edu/admit/freshmen/financial_aid/index.html.

Oliver, Melvin L., and Thomas M. Shapiro. *Black Wealth/White Wealth: A New Perspective on Racial Inequality.* New York: Routledge, 1995.

Olsen, Deborah, and Lizabeth A. Crawford. "A Five-Year Study of Junior Faculty Expectations about Their Work." *The Review of Higher Education* 22.1 (1998): 39–54.

Orfield, Gary, Patricia Marin, and Catherine L. Horn, eds. *Higher Education and the Color Line.* Cambridge: Harvard Education Press, 2005.

Ornstein, Allan C. *Class Counts: Education, Inequality, and the Shrinking Middle Class.* Lanham, MD: Rowman & Littlefield Pub. Group, Inc., 2007.

Osborne, Jason, and Christopher Walker. "Stereotype Threat, Identification with Academics, and Withdrawal from School: Why the Most Successful Students of Colour Might Be Most Likely to Withdraw." *Educational Psychology* 26.4 (2006): 563–577.

Page, Scott E. *The Difference: How the Power of Diversity Creates Better Groups, Firms, Schools, and Societies.* Princeton, NJ: Princeton University Press, 2007.

Palumbo-Liu, David, and Hans Ulrich Gumbrecht. *Streams of Cultural Capital: Transnational Cultural Studies.* Stanford, CA: Stanford University Press, 1997.

Park, Albert, Wen Li, and Sangui Wang. "School Equity in Rural China." *International Conference on Chinese Education.* Teachers College, Columbia University, 2003.

Paulson, William R. *Literary Culture in a World Transformed: A Future for the Humanities.* Ithaca, NY: London: Cornell University Press, 2001.

Plaut, Victoria C. Models of Success in the Academy." *Engaging Our Faculties: Junior Faculty of Color and Senior University Administrators on Diversity and Excellence in Higher Education.* Ed. Stephanie A. Fryberg and Ernesto Javier Martínez (manuscript in progress).

Plaut, Victoria C., and Hazel Rose Markus. "The 'Inside' Story: A Cultural-Historical Analysis of Being Smart and Motivated, American Style." *Handbook of Competence and Motivation.* Ed. Andrew J. Elliot and Carol S. Dweck. New York: Guilford Press, 2007, 457–488.

Podberesky v. Kirwan. 38F.3d 147. 4th Circuit Court of Appeals. 1994.

Pollak, Kathryn I., and Yolanda F. Niemann. "Black and White Tokens in Academia: A Difference of Chronic versus Acute Distinctiveness." *Journal of Applied Social Psychology* 28.11 (1998): 954–972.

Price, Derek V., and Jill K. Wohlford. "Equity in Educational Attainment; Racial, Ethnic, and Gender Inequality in the 50 States." *Higher Education and the Color Line.* Ed. Gary Orfield, Patricia Marin, and Catherine L. Horn. Cambridge: Harvard Education Press, 2005.

Price, Gregory N. "The Problem of the 21st Century: Economics Faculty and the Color Line." *Journal of Socioeconomics* 38.2 (2009): 331–343.

"Professor Protests over Black Admissions at U.C.L.A." *The New York Times.* August 30, 2008.

Quinn, Diane M., and Steven J. Spencer. "The Interference of Stereotype Threat with Women's Generation of Mathematical Problem-Solving Strategies." *Journal of Social Issues* 57.1 (2001): 55–71.

Regents of University of California v. Bakke. 438 U.S. 265. Supreme Court of the United States. 1978.

Rhoads, Robert A., and James R. Valadez. *Democracy, Multiculturalism, and the Community College: A Critical Perspective.* New York: Garland Pub., 1996.

Roberts, Sam. "A Generation Away, Minorities May Become the Majority in U.S." *New York Times.* August 14, 2008: A.1.

Rodgers, W. M., and W. E. Spriggs. "What Does the AFQT Really Measure?: Race, Wages, Schooling and the AFQT Score." *Review of Black Political Economy* 24.4 (June 1996): 13–46.

Rosenthal, Harriet E. S., Richard J. Crisp, and Mein-Woei Suen. "Improving Performance Expectancies in Stereotypic Domains: Task Relevance and the Reduction of Stereotype Threat." *European Journal of Social Psychology* 37.3 (2007): 586–597.

Rothstein, Jesse M. "College Performance Predictions and the SAT." *Journal of Econometrics* 121.1 (2004): 297.

Rudenstine, Neil L. *The President's Report 1993–1995.* Cambridge, MA: Harvard University, 1995.

Ruggles, Steven, Matthew Sobek et al. "Integrated Public Use Microdata Series: Version 3.0." *Minnesota Population Center Data Projects* (2003). http://www.ipums.org.

Sackett, Paul R., Cathy L. DuBois, and Ann W. Noe. "Tokenism in Performance Evaluation: The Effects of Work Group Representation on Male-Female and White-Black Differences in Performance Ratings." *Journal of Applied Psychology* 76.2 (1991): 263–267.

Saenz, Delia S., and Charles G. Lord. "Reversing Roles: A Cognitive Strategy for Undoing Memory Deficits Associated with Token Status." *Journal of Personality and Social Psychology* 56.5 (1989): 698–708.

Salandy, Rassan, ed. *The Posse Foundation 2007 Annual Report.* Ed. The Posse Foundation. Worchester: Saltus Press, 2007.

Grutter v. Bollinger. 2003.

Schmader, Toni, and Michael Johns. "Converging Evidence that Stereotype Threat Reduces Working Memory Capacity." *Journal of Personality and Social Psychology* 85.3 (2003): 440–452.

Schmader, Toni, Brenda Major, and Richard W. Gramzow. "Coping with Ethnic Stereotypes in the Academic Domain: Perceived Injustice and Psychological Disengagement." *Journal of Social Issues* 57.1 (2001): 93–111.

Schmidt, Peter. "Cold Reality Intrudes on Diversity Conference in Disney World." *Chronicle of Higher Education.* May 30, 2008. http://chronicle.com/article/Cold-Reality-Intrudes-on/849 (accessed February 5, 2010).

———. *Color and Money: How Rich White Kids Are Winning the War over College Affirmative Action.* 1st ed. New York: Palgrave Macmillan, 2007.

Schrodt, Paul, Carol Stringer Cawyer, and Renee Sanders. "An Examination of Academic Mentoring Behaviors and New Faculty Members Satisfaction with Socialization and Tenure and Promotion Processes." *Communication Education* 52.1 (2003): 17–30.

Sekaquaptewa, Denise. "On Being the Solo Faculty Member of Color: Research Evidence from Field and Laboratory Studies." *Engaging Our Faculties: Junior Faculty of Color and Senior University Administrators on*

Diversity and Excellence in Higher Education. Ed. Stephanie A. Fryberg and Ernesto Javier Martínez (manuscript in progress).

Sekaquaptewa, Denise, and Mischa Thompson. "Solo Status, Stereotype Threat, and Performance Expectancies: Their Effects on Women's Performance." *Journal of Experimental Social Psychology* 39.1 (2003): 68–74.

Sen, Amartya. *On Ethics & Economics*. Cambridge: Blackwell, 1994.

Shapiro, Stewart, and Mark T. Spence. "Mind over Matter? The Inability to Counteract Contrast Effects Despite Conscious Effort." *Psychology and Marketing* 22.3 (2005): 225–245.

Shaw, Kathleen M., James R. Valadez, and Robert A. Rhoads, eds. *Community Colleges as Cultural Texts: Qualitative Explorations of Organizational and Student Culture*. Albany, NY: State University of New York Press, 1999.

Shore, Bradd. *Culture in Mind: Cognition, Culture, and the Problem of Meaning*. New York: Oxford University Press, 1996.

———. *What Culture Means, How Culture Means*. Worchester, MA: Clark University Press, 1998.

Smith, D., & Associates. *Diversity Works: The Emerging Picture of How Students Benefit*. Washington, DC: Association of American Colleges and Universities, 1997.

Smith, Jessi L., Carol Sansone, and Paul H. White. "The Stereotyped Task Engagement Process: The Role of Interest and Achievement Motivation." *Journal of Educational Psychology* 99.1 (2007): 99–114.

Smith, Mathew A., Sandy Baum, and Michael S. McPherson. "Financial Independence and Age: Distributive Justice in the Case of Adult Education." *Theory and Research in Education* 6.2 (June 2008): 131–152.

"Special Report: Ever Higher Society, Ever Harder to Ascend—Meritocracy in America." *The Economist* 374.8407 (2005): 35.

Spencer, Steven J., Claude M. Steele, and Diane M. Quinn. "Stereotype Threat and Women's Math Performance." *Journal of Experimental Social Psychology* 35.1 (1999): 4–28.

Sperber, Dan. *On Anthropological Knowledge: Three Essays*. Cambridge [Cambridgeshire]; New York: Cambridge University Press, 1985.

Spindler, George Dearborn, and Louise S. Spindler. *The American Cultural Dialogue and Its Transmission*. London; New York: Falmer Press, 1990.

"Stanford Enhances Undergraduate Financial Aid Program." *Stanford Report* (February 20, 2008). http://news-service.stanford.edu/news/2008/february20/finaid-022008.html (accessed February 21, 2008).

"Stanford Medical Youth Science Program (SMYSP) 'Evaluation Results.'" *Stanford University School of Medicine*. http://smysp.stanford.edu/evaluationResults/.

"Stanford Medical Youth Science Program (SMYSP) 'Mission.'" *Stanford University School of Medicine*. http://smysp.stanford.edu/about/.

Stangor, Charles, Christine Carr, and Lisa Kiang. "Activating Stereotypes Undermines Task Performance Expectations." *Journal of Personality and Social Psychology* 75.5 (1998): 1191–1197.

Steele, Claude M. *Expert Report, Gratz and Grutter.* U.S. District for the Eastern District of Michigan.
Steele, Claude M. "A Threat in the Air: How Stereotypes Shape Intellectual Identity and Performance." *American Psychologist* 52.6 (1997): 613–629.
Steele, Claude M., Steven J. Spencer, and Joshua Aronson. "Contending with Group Image: The Psychology of Stereotype and Social Identity Threat." *Advances in Experimental Social Psychology* 34 (2002): 379–440.
———. "Stereotype Threat and the Intellectual Test Performance of African Americans." *Journal of personality and social psychology* 69.5 (1995): 797–811.
"Steering Committee on Slavery and Justice." *Brown University.* http://www.brown.edu/Research/Slavery_Justice/.
Stone, Jeff. "Battling Doubt by Avoiding Practice: The Effects of Stereotype Threat on Self-Handicapping in White Athletes." *Personality and Social Psychology Bulletin* 28.12 (2002): 1667–1678.
Student Academic Services, Office of the President. *Undergraduate Access to the University of California after the Elimination of Race-Conscious Policies,* March 2003.
Sugrue, Thomas J. *The Origins of the Urban Crisis [Electronic Resource]: Race and Inequality in Postwar Detroit.* Ed. American Council of Learned Societies. Princeton, NJ: Princeton University Press, 1996.
Swan, Suzanne, and Robert S. Wyer, Jr. "Gender Stereotypes and Social Identity: How Being in the Minority Affects Judgment of Self and Others." *Personality and Social Psychology Bulletin* 23.12 (1997): 1265–1276.
Swim, Janet K., and Deborah L. Miller. "White Guilt: Its Antecedents and Consequences for Attitudes toward Affirmative Action." *Personality and Social Psychology Bulletin* 25.4 (1999): 500–514.
Tatum, Beverly Daniel. *"Why Are All the Black Kids Sitting Together in the Cafeteria?": A Psychologist Explains the Development of Racial Identity.* New York: Basic Books, 1997.
Taylor, Shelley E., Susan T. Fiske, Nancy L. Etcoff, and Audrey J. Ruderman. "Categorical and Contextual Bases of Person Memory and Stereotyping." *Journal of Personality and Social Psychology* 36.7 (1978): 778–793.
Thoman, Dustin B., Paul H. White, Niwako Yamawaki, and Hirofumi Koishi. "Variations of Gender-Math Stereotype Content Affect Women's Vulnerability to Stereotype Threat." *Sex Roles* 58.9–10 (2008): 702–712.
Thompson, Dennis F. "The Moral Responsibility of Many Hands." *Political Ethics and Public Office.* Cambridge, MA: Harvard University Press, 1987, 40–65.
Thompson, M. C., T. G. Brett, and C. Behling. "Educating for Social Justice: The Program on Intergroup Relations, Conflict, and Community at the University of Michigan." *Intergroup Dialogue: Deliberative Democracy in School, College, Community, and Workplace.* Ed. David Louis Schoem and Sylvia Hurtado. Ann Arbor, MI: University of Michigan Press, 2001, 99–114.

Thompson, Mischa, and Denise Sekaquaptewa. "When Being Different Is Detrimental: Solo Status and the Performance of Women and Racial Minorities." *Analyses of Social Issues & Public Policy* 2.1 (2002): 183–203.

Tilghman, Shirley M. "Changing the Demographics: Recruiting, Retaining, and Advancing Women Scientists in Academia." Earth Institute ADVANCE Program. Columbia University, March 24, 2005.

Tippeconnic, John W., and Smokey McKinney. "Native Faculty: Scholarship and Development." *The Renaissance of American Indian Higher Education: Capturing the Dream*. Ed. Maenette K. P. AhNee-Benham and Wayne J. Stein. Mahwah, NJ: Lawrence Erlbaum, 2003, 241–296.

Tomasello, Michael. *The Cultural Origins of Human Cognition*. Cambridge, MA: Harvard University Press, 1999.

Trounson, R. "Scholarship Fund to Help Blacks Go to UCLA." *Los Angeles Times*. March 29, 2007.

Turner, Caroline Sotello Viernes, and Samuel L. Myers. *Faculty of Color in Academe: Bittersweet Success*. Boston, MA: Allyn and Bacon, 2000.

U.S. Census Bureau. *Demographic Trends in the 20th Century*. Washington, DC: U.S. Department of Commerce, 2002.

U.S. Census Bureau. Housing and Household Economics Statistics Division. *Racial and Ethnic Residential Segregation in the United States: 1980–2000*, August 11, 2008.

U.S. Department of Education. *Appendix 5: U.S. Department of Education Final Policy Guidance, Nondiscrimination in Federally Assisted Programs; Title VI of the Civil Rights Act of 1964*. Vol. 59, No. 36., February 23, 1994.

———. *Digest of Educational Statistics, 2002*. Report No. NCES 2003–060, Table 133. National Center for Education Statistics: Office of Educational Research and Improvement, Washington D.C.

———. National Center for Education Statistics. http://nces.ed.gov/college navigator/?q=harvard&s=all&id=166027 (accessed February 14, 2008).

———. "The Use of Race in Postsecondary Student Admissions" (August 28, 2008). www.ed.gov/about/offices/ocr/letters/raceadmissionspse.html.

"The University of Michigan, Ann Arbor: 2003 Freshman Class Profile" (2003). http://www.umich.edu.proxy.library.cornell.edu/~oapainfo/TABLES/FR_Prof.html#REB.

Villalpando, Octavio, and Dolores Delgado Bernal. "A Critical Race Theory Analysis of Barriers that Impede Success of Faculty of Color." *The Racial Crisis in American Higher Education: Continuing Challenges for the Twenty-First Century*. Ed. William A. Smith, Philip G. Altbach, and Kofi Lomotey. Albany, NY: State University of New York Press, 2002, 243–270.

Wagner, Sally Roesch. *Sisters in Spirit: The Iroquois Influence on Early American Feminists*. Summertown, TN: Native Voices, 2001.

Wechsler, Harold S. *The Qualified Student: A History of Selective College Admission in America*. New York: Wiley, 1977.

Wellesley College. "The College: An Introduction." http://www.wellesley.edu/Welcome/college.html.
Wiedeman, R. "Analysis: How Colorado Became the First State to Reject a Ban on Affirmative Action." *The Chronicle of Higher Education* (November 10, 2008). http://chronicle.com/article/Analysis-Why-Colorado-Failed/1317 (accessed February 5, 2010).
Wilgoren, Jodi. "In One Prison, Murder, Betrayal and High Prose." *New York Times.* April 29, 2005: A.16.
"The Will of George Washington (July 9, 1799)." *University of Virginia.* http://www.gwpapers.Virginia.edu/documents/will/text.html.
Williams, John B. *Race Discrimination in Public Higher Education: Interpreting Federal Civil Rights Enforcement, 1964–1996.* Westport, CT: Praeger, 1997.
Wilson, William J., and Richard P. Taub. *There Goes the Neighborhood: Racial, Ethnic, and Class Tensions in Four Chicago Neighborhoods and Their Meaning for America.* New York: Knopf, 2006.
Witte, G., and N. Henderson. "Wealth Gap Widens for Blacks, Hispanics." *The Washington Post.* October 18, 2004.
Yoder, Janice D. "Making Leadership Work More Effectively for Women." *Journal of Social Issues* 57.4 (2001): 815–828.
———. "Rethinking Tokenism: Looking Beyond Numbers." *Gender and Society* 5.2 (1991): 178–192.
Yoder, Janice D., Thomas L. Schleicher, and Theodore W. McDonald. "Empowering Token Women Leaders: The Importance of Organizationally Legitimated Credibility."" *Psychology of Women Quarterly* 22.2 (1998): 209–222.
Zárate, Michael A., and Eliot R. Smith. "Person Categorization and Stereotyping." *Social Cognition* 8 (1990): 161–185.
Zernike, Kate, and Dalia Sussman. "For Pollsters, the Racial Effect that Wasn't." *New York Times.* November 6, 2008: P.8.
Zusman, Ami. "Issues Facing Higher Education in the Twenty-First Century." *American Higher Education in the Twenty-First Century: Social, Political and Economic Challenges.* Ed. Philip Altbach, Robert Berdahl, and Patricia Gumport. Baltimore, MD: Johns Hopkins University Press, 1999, 109–150.

Contributors

Gregory M. Anderson is the Dean of the Morgridge College of Education at the University of Denver and a tenured Associate Professor in Education. Before coming to DU in 2009, Dr. Anderson was an Associate Professor at Columbia University's Teachers College Program in Higher and Postsecondary Education. In 2006, Anderson was granted an extended leave from Teachers College to become the higher education policy program officer for the Ford Foundation in New York. He was responsible for overseeing one of the largest portfolios at the Foundation featuring both international and domestic higher education grants. Anderson also sat on executive committees of multifoundation partnerships and foundation-wide initiatives involving the United States, Africa, Central and Latin America, and Asia. In 2008, he was appointed by the Vice President of the Foundation's Knowledge, Creativity and Freedom Program Division to lead a strategic planning team responsible for developing a new vision for the United States and international higher education programming. Anderson earned a PhD in sociology from the Graduate Center at the City University of New York and is currently a member of the editorial board for the *Review of Higher Education*.

Nancy Cantor, Chancellor of Syracuse University and former Chancellor of the University of Illinois at Urbana-Champaign, has worked to forge new understandings of the university as a public good, promoting community engagement, racial justice and diversity, the status of women, the creative campus, and sustainability. As Provost of the University of Michigan, she was closely involved in the university's defense of affirmative action before the Supreme Court. A social psychologist, she is the author of numerous books, chapters, and scientific articles. She is a fellow of the American Academy of Arts and Sciences and a member of the Institute of Medicine of the National Academy of Sciences.

William A. ("Sandy") Darity, Jr. is Arts and Sciences Professor of Public Policy Studies and Professor of African and African American

Studies and Economics at Duke University. He previously served as Director of the Institute of African American Research, the Moore Undergraduate Research Apprenticeship Program, the Undergraduate Honors Program in economics, and Graduate Studies at the University of North Carolina. He was a fellow at the National Humanities Center (1989–90) and a visiting scholar at the Federal Reserve's Board of Governors (1984). He is a past president of the National Economic Association and the Southern Economic Association. He has also taught at Grinnell College, the University of Maryland at College Park, the University of Texas at Austin, Simmons College, and Claremont-McKenna College. He recently served as the Editor in Chief of the new edition of the *International Encyclopedia of the Social Sciences* (Macmillan Reference 2008); he is the author of *Economics, Economists, and Expectations: Microfoundations to Macroapplications* (2004) (coauthored with Warren Young and Robert Leeson), and coeditor with Ashwini Deshpande of *Boundaries of Clan and Color: Transnational Comparisons of Inter-Group Disparity* (2003), both published by Routledge. He has published or edited 10 books and more than 200 articles in professional journals.

Steven J. Diner has served as Chancellor of Rutgers University-Newark since 2002, and as Dean of Arts and Sciences from 1998 to 2002. He is also Professor of History. Before coming to Rutgers-Newark, Diner served as Vice Provost, Associate Senior Vice President and Professor of History at George Mason University and as a faculty member in Urban Studies and Chair of the Urban Studies Department at the University of the District of Columbia. He currently serves as President of the Coalition of Urban and Metropolitan Universities. His publications include *A City and Its Universities: Public Policy in Chicago, 1892–1919* (1980) and *A Very Different Age: Americans of the Progressive Era* (1998).

Stephanie A. Fryberg is an Assistant Professor in the Department of Psychology and an Affiliate Faculty member in American Indian Studies at the University of Arizona. Her primary research interests focus on how social representations of race, culture, and social class influence the development of self. Her recent publications include *Of Warrior Chiefs and Indian Princesses: The Psychological Consequences of American Indian Mascots on American Indians* (with H. R. Markus, D. Oyserman, and J. M. Stone), *Identity-Based Motivation and Health* (with D. Oyserman and N. Yoder), *Cultural Models of Education in American Indian, Asian American, and European American Contexts* (with H. R. Markus), and *The Psychology of*

Engagement with Indigenous Identities: A Cultural Perspective (with G. Adams, D. M. Garcia, and E. U. Delgado). In 2007, Dr. Fryberg was the recipient of the Society for the Psychological Study of Social Issues (SPSSI) Louise Kidder Early Career Award for contributions of research to society and the University of Arizona Five Star Faculty Award for excellence in undergraduate education.

Michael Hames-García has been Program Director and Department Head of Ethnic Studies at the University of Oregon since 2006. He also directs the Center for Race, Ethnicity, and Sexuality Studies (CRESS) at the University of Oregon. He is a founding member of the Future of Minority Studies Research Project (FMS) and the author and editor of several books, including *Reclaiming Identity: Realist Theory and the Predicament of Postmodernism* (University of California Press 2000), *Fugitive Thought: Prison Movements, Race, and the Meaning of Justice* (University of Minnesota Press 2004), and *Identity Politics Reconsidered* (Palgrave 2006).

Muriel A. Howard is President of the American Association of State Colleges and Universities (AASCU), a post she has held since August 2009. Prior to assuming leadership of AASCU, Dr. Howard was President of Buffalo State College (State University of New York) from 1996 to 2009. Her professional and scholarly interests include support of education, educational leadership, representation of women and minorities, and public service. As AASCU President Dr. Howard advocates for public higher education at the national level, working to influence federal policy and serving as a resource for presidents and chancellors at 430 campuses nationwide.

Marvin Krislov became the fourteenth President of Oberlin College in 2007, where he is also a professor in the Politics Department. Mr. Krislov served as Vice President and General Counsel at the University of Michigan from 1998 to 2007, where he was also an adjunct professor for the Law School and the Political Science Department. He led the University of Michigan's legal defense of its admissions policies, resulting in the 2003 Supreme Court decision recognizing the importance of student body diversity.

Jeffrey S. Lehman is Professor of Law and former President of Cornell University, currently on leave to be Chancellor and Founding Dean of the Peking University School of Transnational Law. He previously served as Dean of the University of Michigan Law School, where he was one of the architects of the school's successful defense of its admissions program before the U.S. Supreme Court. His scholarship deals with the poverty and inequality in the American welfare state; his most recent book is *An Optimistic Heart: What Great Universities Can Give Their Students...and the World*.

Daniel Little has served as Chancellor of the University of Michigan-Dearborn since 2000, where he is also a professor of philosophy. Previously he served as Vice President for Academic Affairs at Bucknell University and as Associate Dean of Faculty at Colgate University. He is a philosopher of social science, with a continuing interest in the foundations of sociology. His most recent book, *The Paradox of Wealth and Poverty: Mapping the Ethical Dilemmas of Global Development*, is a study of the ethics of international development.

Michael S. McPherson is President of the Spencer Foundation. Prior to joining Spencer he served as President of Macalester College in St. Paul, Minnesota, for seven years. He is a nationally known economist whose expertise focuses on the interplay between education and economics. McPherson, who is coauthor and editor of several books, including *College Access: Opportunity or Privilege?*, *Keeping College Affordable*, and *Economic Analysis and Moral Philosophy*, was founding coeditor of the journal *Economics and Philosophy*.

Satya P. Mohanty has taught at Cornell since 1983, where he is currently Professor of English. He is one of the founders of the FMS Research Project (2000–) and has been the Director of the national FMS Summer Institute since 2005. His scholarly work, including his book *Literary Theory and the Claims of History* (1997), deals in part with the relationship between minority identities and social justice.

Matthew A. Smith is a 2012 J.D. candidate at the Yale Law School. He graduated Phi Beta Kappa with honors and with distinction from Stanford University in 2006, where he was awarded the Golden Medal for Excellence in Humanities and the Rheinlander Prize for Outstanding Work in Philosophy. He is a Point Foundation Scholar and worked for several years as a research associate at the Spencer Foundation in Chicago, Illinois. In that capacity he coauthored numerous scholarly works on ethical issues in higher education.

Teresa A. Sullivan has been Provost and Executive Vice President for Academic Affairs at the University of Michigan since 2006 and in August 2010 will become President of the University of Virginia. Previously she was Executive Vice Chancellor of the University of Texas System. She is a sociologist and demographer and served as Co-Principal Investigator of the Texas Higher Education Opportunity

Project, a multiyear study of the effects of the Texas Top Ten Percent law on college admissions.

Eugene M. Tobin, a former President of Hamilton College, is Program Officer for higher education and the liberal arts colleges at The Andrew W. Mellon Foundation, and is the coauthor of *Equity and Excellence in American Higher Education*. His most recent essay, "The Modern Evolution of America's Flagship Universities," will appear in William G. Bowen, Matthew M. Chingos, and Michael S. McPherson's *Crossing the Finish Line: Completing College at America's Public Universities* (Princeton University Press 2009).

Index

academic
  climate, 15
  freedom, 81
  preparation, 57, 62, 99, 100, 111, 112, 119
  segregation, 174
  standards, 134
"Academic Bill of Rights," 28
academy, 186, 189, 190, 193–194, 198, 204–206
  culture of, 182–184
  demographics of, 208
  diversification of, 184
  racial-ethnic composition of, 183
access, 4, 20, 77, 94, 99, 140, 167
  for African Americans, 133
  and equity, 77, 82
  and excellence, 76–77
  to higher education, 10, 12, 75, 111, 127, 138, 141
    barriers to, 98
    outcomes of, 75
  inequalities of, 69, 79, 81, 109
  rates of minority, 131
ACT scores, 73
"actively integrated pod," 45–46
activism, 32
  student, 6, 52, 64
admission(s), 6, 11, 100–102, 115, 132–133, 137–139, 147, 151, 155, 161, 163, 167
  criteria, 11–12, 131, 138
  gender-conscious, 165
  need-blind, 102–103, 115
  policies, 103
    race-neutral, 131, 162
  preferences in, 102, 103, 114, 161
    athletics, 101–103, 114
    class-based, 170
    income-based, 10, 104, 137
    income-sensitive, 104
    legacy, 101–102, 114
    minority, 101–103, 114
    race-sensitive, 104–106, 131, 133, 170
  race-conscious, 103, 126, 131, 163
  use of race in, 148, 151, 161, 162
affirmative action, 4, 10, 13, 32, 38, 42–43, 50, 52, 87, 99, 102, 104, 126–127, 130–131, 133, 137, 139–140, 148, 151, 154, 156–157, 159–161, 163, 165, 168
  anti-, 105
  ban(s) on, 148, 151, 155, 163, 164
  benefits of, 132
  definition of, 169
  economic, 163
Alcoff, Linda Martín, 192–194
American Association of State Colleges and Universities (AASCU), 74, 89, 91
American Economic Association, 14, 174, 176–177
  journals, 14, 174–176
American Indian Center, 22
American Indian Studies, 13, 181, 185
Anderson, Gregory M., 11, 123, 239

Andrew W. Mellon Foundation, 10, 97
Antonio, Anthony Lising, 64–65, 67
"apartheid public schools," 19
Aronson, Joshua, 1, 196
arts, 24, 35–36
Association of American Universities (AAU), 84
Attewell, Paul, 75

Bane, Mary Jo, 47
Beijing, 78–82
Bennett, Bill, 76
black economists, 174–175
black scholars, 14, 173, 174, 177, 178, 179
  contributions of, 14, 179
black-white binary, 192
black-white gap, 133, 177
Bok, Derek, 105, 125, 131–135, 141
Bowen, William, 10, 44, 56, 58, 59, 62, 77, 97, 105, 125, 131–135, 141, 163
bridge programs, 72
Brown University, 55, 61–62, 103, 174
  Steering Committee on Slavery and Justice, 62, 66
Brown v. Board of Education, 58
Bucknell University, 69
Buffalo State College, 8, 89–96
  Students' Awards, 93
Business-Higher Education Forum, 99

California, 9
  Hi-Voltage decision, 165
  Prop 209, 165
campus
  administration, 9
  diverse, 4, 5, 87, 209
  inclusive, 4, 14, 168
  integrated, 5, 42, 43, 45
  leadership, 9
  multicultural, 87

campus cafeteria, 5, 43, 44, 46, 48
campus culture, *see under* culture
Canada, 94
  model of multiculturalism, 94–95
Cantor, Nancy, 4, 6, 38, 66, 239
Center for Race, Ethnicity, and Sexuality Studies, 51
Chief Illiniwek, 21–22
China, People's Republic of, 70, 78–81
  admissions system, 79–80, 81
  education system, 79–80, 81
Chronicle of Higher Education, 53
City University of New York (CUNY), 8, 75
Civil Rights Act of 1964, 128
Civil Rights Project, 36
  Harvard, 19, 35
class
  opportunity gap, 135
  socioeconomic, 10
  stratification, 88
Clinton, Hillary, 159–160
Colgate University, 69
College and Beyond database, 131
College Board, 85, 100, 134, 155
college preparation, 10–11, 71–72, 91, 99, 100, 111–113, 118–119, 168
  programs, 72
color-blind, 16, 23, 45
Columbia University, 11, 123
commodity, 125–126
communities of color, 210
community colleges, 9, 85, 135
conflict, 23, 24
  consciousness of, 5
  intercultural, 19
  normalizing, 5, 28
Congress, 10, 113, 118
  U.S. House of Representatives, 113
Connective Corridor, 35
Constitution, 148, 161, 163
Cornell University, 1, 5, 41, 49, 50, 54–55
  joint student programming at, 49

critical feedback, 16, 17
critical mass, 130
  achieving, 52, 53, 104, 128, 162, 163
cross-cultural exchange, 166
cues, 1
cultural
  capital, 140
  competence, 61, 185
  dominance, 27
  dynamics, 87
  identity, 95, 129
  inter-
    conflicts, 28
    experiences, 24
culture, 3
  campus, 5, 15
    democratic, 27
    effective, 15
    ideal, 6
    impact of faculty on, 5
    negative, 14
  democratic, 5, 20
  diverse, 5, 6
  of learning, 4
curriculum
  internationalization of, 126

Darity, Jr., William A., 13, 173, 239–240
defensive barriers, 31
democracy, 29, 42, 123–124, 130
  Amerian, 76
democratic
  community, 19
  culture, 5, 20, 21, 23, 26, 27, 32, 34–36
  outcomes, 130
  society, 5, 9
demographic landscape, 19
demographic shift, 126
demographics of diversity, 99
Descriptor PLUS, 154, 155
Detroit, 32, 70, 71
  Public Schools, 71, 73
  unemployment in, 71

dialogue, 19, 28
  difficult, 24, 29, 32
  on diversity, 29
  honest, 56
  safe, 31
difference, 4, 20, 23, 53, 56, 58, 96, 136
  appreciating, 26
  multicultural, 90, 104, 110
  tolerance of, 6
  "differentness," 149
Diner, Steven J., 7–8, 83, 240
disadvantaged backgrounds, 140
discrimination, 87, 124, 148, 150, 151, 159–161
discriminatory outcomes, 14
dissimilarity index, 70
diversity, 3–4, 8, 10, 15, 17, 26, 35, 51, 53, 58–59, 61, 62, 65, 87, 90, 92, 94, 97–99, 103, 126, 128, 134–135, 138–141, 147–148, 156, 160, 166–168, 185, 204, 208–209
  benefits of, 125–127, 134–136, 148, 156
  campus, 43, 167
  celebration of, 61, 124
  as compelling interest, 161
  consumption of, 11
  creation of, 19
  cultural, 6
  discourse of, 126, 136
    hegemonic, 127
  economic, 165
  educational benefits of, 27, 105, 128–129, 132
  future of, 17
  geographic, 149
  ideological, 175
  initiatives, 8, 10, 15, 52, 60, 207–208
  intellectual, 19
  and leadership, 129, 161
  measuring, 8
  multicultural, 72

diversity—*Continued*
  officers/offices of, 6, 53–55
  racial, 13, 59, 98, 105, 165
  requirements, 6, 60, 166
  social, 4, 6, 13
    as resource, 17
  sociocultural, 7
  value of, 26, 136
divided society, 29
"dominant" group, 28
"double consciousness," 205
Duke University, 13, 173

economic
  benefits of education, 118
  development, 9
  downturn, 9, 140, 163, 167
economically disadvantaged, 73, 114
economics
  departments, 13
    diversification of, 174
    midnight, 173, 174
    sundown, 173
    window dressing, 173, 174
  field of, 14, 179
  and race, 174–175
  journals, 176–178
    "general," 176, 178
    "open," 176
    Small vs. Big, 177–178
    "specialty," 176
economy, 89
  engine-, 89–92, 94, 96
  knowledge-based, 96
education
  affordable mass, 75
educational
  advantages, 109
  attainment, rate of, 99
  background, 12
  excellence, 7–9, 15, 77
  experience, 9
  interest, 161
  opportunity, 71
  policy, 2, 10, 11
  psychology, 4
  quality, 9
  resource, 4, 8, 27
educational equity, 192, 194
Educational Opportunity Fund (EOF), 85
Ellwood, David, 47
empathy, 23, 26, 28, 29
environment
  integrated, 45, 48
  learning, 4, 5, 7, 16
    democratic, 130
    diverse, 129, 131
    inclusive, 16, 17, 208
    optimal, 167
  "majority minority," 44
  testing, 16
Equal Protection Clause, 41, 42
equality, 150
equality of opportunity, 70, 76
equity and diversity, 51
  offices of, 57, 61
equity and excellence, 73, 74, 78, 98, 103
Equity and Excellence, 10, 18, 97, 106, 110
ethnic studies
  department, 6, 51, 53
  program, 52, 64
exclusion, 90, 94

faculty
  of color, 15, 30, 51, 52, 55, 56, 65, 160, 196
    junior, 14–15, 182–186, 189
  experience with diversity, 30
  junior, 30, 31, 52, 183, 187, 190–191, 199
    of color, 193–196, 198–200, 202–206, 208–210
    solo status of, 198–201
  mentoring, 30–31
  senior, 30, 31, 182–183, 190, 199, 207–208
    of color, 193
    as cultural gatekeepers, 183

INDEX 249

"successful," 183–184
white, 65
fetishism, 124
fetishization
  of diversity, 14, 141
  of merit, 14, 179
financial aid, 9, 104, 112–115, 120, 139, 140, 163–165, 167
  federal and state, 116–118, 120
  merit vs. need-based, 165
  policies, 119, 167
  policy makers, 112–113
  as roadblocks, 116–117
  targeted, 154, 155, 164, 166
first-generation
  college students, 7, 101–102, 114, 138, 163
  immigrant families, 7
Ford Foundation, 11, 123
Free Application for Federal Student Aid (FAFSA), 118–119
Fryberg, Stephanie A., 13–14, 181, 240–241
Future of Minority Studies (FMS) Research Project, 1, 51, 185, 192, 194

gap
  opportunity, 12
  rich-poor, 11
gatekeepers, 13
gay/lesbian, 7, 63
gender-blind, 16
geographic
  diversity, 161
  representation, 153
geography, 97, 147, 149, 155–156
  and access, 153–156
  macro-, 149–151
  micro-, 151–153, 154, 156
George Mason University, 83
Gerken, LouAnn, 190–191
Graduate Record Examinations (GRE), 1, 196
Gratz v. Bollinger, 13, 59, 126, 128–129, 162

group(s)
  -based inequalities, 29
  dynamics, 4, 20
  identities, 5, 21
  (in), 23
  inter-
    appreciation, 23
    conflict(s), 26, 27, 31
    defenses, 26
    dialogue, 23–24, 26–27, 31, 35, 166
    differences, 31
    dynamics, 27
    exchange, 5, 20–22, 26, 32
    experiences, 24
    mentoring, 31
    relations, 20–22, 29, 36, 150
      curriculum, 27, 35
    tensions, 29, 30
  intra-
    affirmation, 23, 31, 32
    consensus, 31
    security, 27
  minority, 23, 44
  (out), 23, 26
  socially marginalized, 5, 28, 29
Grutter v. Bollinger, 5, 13, 41–43, 59, 104, 126, 128–129, 143, 151, 154, 161, 162, 164, 166
Gurin, Patricia, 23, 27, 29, 31, 37, 98, 105, 129, 130
  research on University of Michigan, 129–130, 143
Gutmann, Amy, 97

Hames-García, Michael, 4, 6, 51, 241
Hamilton College, 10, 97
Harvard University, 19, 54, 55, 103, 105, 112, 114–116, 136, 160, 162, 170, 174, 176
hate crimes, 63
Haudenosaunee, 33
  Chief Sidney Hill, of, 33–34
  Great Law of Peace, 33
  Promise program, 34, 61, 66
  protections to women, 33

higher education, 83, 100
  benefits of, 75
  as a commodity, 99
  declining state support for, 76, 83
  diversity in, 98, 124, 136, 140
    as a commodity, 125, 137
  diversity initiatives, 60
  equity in, 58, 70
  excellence in, 58, 70
  opportunity gap in, 141
  public funding for, 77, 83
  rising costs of, 167
  stratification of, 85
Higher Education Opportunity Program (HEOP), 145
historically black colleges and universities, 62
Hopwood v. Texas, 148, 153–155, 161
Hosseini, Khalid, 24–26, 37
Howard, Muriel A., 8, 89, 241
Hu-DeHart, Evelyn, 53
human potential, 10
human talent, 82

identity
  academic, 185
  issues, 6
  minority, 185, 192
immigrant(s), 8, 9, 84, 86, 95, 148, 149, 156
  culture and identity of, 86
  family, 7, 94
immigration, 126
immigration patterns, 141
income
  low-, lower-, 9
    backgrounds, 60, 84, 102, 136
    eligibility reuiqrements, 138
    families, 101, 103, 110, 116, 117, 121
    policy, 137
    status, 10
    students, 11, 100, 115, 118–120

individualism, 202–205
inequality, 4, 10, 56, 57, 69–71, 79, 120, 124, 205
  in Detroit, 70
  educational, 11, 70–71, 109–113, 118–120
  income, 70
  racial, 70, 178
  social, 6, 9, 10, 61, 62
  socioeconomic, 11
  structural, 6, 27–28, 53, 58, 60, 66
  systemic, 204
Inner Mongolia, 78, 80
institutional bias, 195
institutions
  elite, 7, 12, 76, 77, 115, 125, 132, 133, 135, 138, 141, 160
  national, 7
  nonelite, 73, 76, 82
  nonflagship, 7–8, 74, 82
  regional, 7–9, 73, 76, 77, 82
  state, 9, 83
  urban, 8, 9
integration, 44, 48, 50, 94
  cross-racial, 5
integration pods, 5, 6, 46, 47, 49
interaction(s), 5, 17, 28, 98, 130, 166, 190, 199
  cross-racial, 136
  healthy, 5, 27
Ivy League, 6, 52

Keats, 44
The Kite Runner, 24–26, 37
Krislov, Marvin, 12–13, 159, 241
Kristof, Nicholas, 119
Krugman, Paul, 9–10
Kurzweil, Martin, 10, 44, 58, 59, 62, 97, 114

Latino(a), 7, 63
Lavin, David, 75
Lehman, Jeffrey S., 4–6, 41, 241–242

lesbian, gay, bisexual, transgender (LGBT)
  community, 28
  studies, 28
liberal arts colleges, 114
Little, Daniel, 7–8, 66, 69, 242
Locke, John, 202–203
Loury, Glenn, 99, 104

Macalester College, 11, 109
Macdonald, Amie, 63, 67
Mann, Charles C., 33
"many hands," 113
Markus, Hazel Rose, 188–189, 197, 216, 217
Marx, Karl, 125, 141
McPherson, Michael, 11, 109, 242
Mellon Mays Undergraduate Fellowship (MMUF), 59
mentorship, 15, 193, 204, 208
merit, 13–15, 115, 124, 138, 148, 203
meritocracy, -tic, 98, 163, 202–205
Michigan, 77
Michigan consortium, 29
Milliken v. Bradley, 153
minority, -ies, 104, 125–126, 137, 152, 159, 160, 163, 166
  identity, 44, 185, 192
  student(s), 101, 134, 170
mobility, 19
models of success, 15
Mohanty, Satya P., 1, 66, 242
multicultural
  affairs, 6, 52, 55, 56
    offices of, 54–57, 62, 166
    see also equity and diversity, offices of
  difference, 90
multiculturalism, 43, 58, 60, 94, 96, 126
Multiversity Intergroup Dialogue Project, 28, 29

National Collegiate Athletic Association, 22

National Educational Longitudinal Study, 100, 108, 126
Native American
  communities, 62, 186–187
  students, 60, 61, 74, 86, 104, 187
  studies, 34, 61, 185, 187
"negative capability," 44
negative social images, 1
neutrality, 15
  blind, 16
New York State, 25, 93
New York Times, 9, 24, 33

Obama, Barack, 123–124, 136, 159–160, 168
Oberlin College, 12, 159, 165, 170
objectivity, 15, 16, 17
O'Connor, Sandra Day, 41, 42, 43, 105
Onondaga Nation, 33, 34
opportunity gap, 141
The Origins of the Urban Crisis, 71

Passing the Torch, 75
Peabody Individual Achievement Test, 113
pedagogy, -ies, 3, 7, 9, 15, 76, 119, 184, 208
  experimentation, 8
Peking University (PKU), 41, 78–80
Pell Grant, 113, 146
perspective enlargement, 45, 47, 50
pipeline, 163, 167, 174, 181
pluralistic society, 29
Podberesky v. Kirwan, 164
policy,
  admissions, 11
  economic, 9
  educational, 2, 10, 11
  national, 10
  social, 9
  socioeconomic, 11

Posse Foundation, 11, 139, 165
  Dynamic Assessment Process, 11, 139
poverty, 47–48, 114, 119, 120
  childhood, 11, 119
  and cognitive gaps, 119–120
  in Detroit, 71
  and higher education, 75–76
  rural, 139
  spells
    dynamics of, 47–48
    stresses of, 119
Powell, Lewis, 128, 162
predominantly white
  campus(es), 134
  student bodies, 136, 152
prejudice(s), 15, 90, 148, 151, 187
Princeton University, 103, 146, 174, 181
privilege, 29, 99
public
  education, 72
  good, 19, 34
  interest, 99
  universities, 155
    funding of, 9

Questbridge, 165

race, 56, 70, 72, 104, 123, 124, 150, 159, 201, 209
  -based programs, 10
  category of, 43, 49
  and class, 127, 134, 137
  class basis of, 11
  -conscious aid, 164
  -conscious perspective, 60
  -conscious policies, 127, 130, 131
  as a criterion, 59
  disadvantages of, 69
  and higher education, 57
  relations, 104
  as social construct, 150
racial
  barriers, 71, 123
  bi-, 49
  composition, 10
  disadvantage(s), 69, 71–73
    urban, 72
  discrimination, 124, 132, 136, 178
  diversity, 13, 59, 98, 105, 165
  equality, 63
  exclusion, 174
  inclusion, 160
  inequality, 123, 178
  injustice, 61
  integration, 58
  justice, 69, 72
  labels, 149
  lenses, 16
  minority, 99, 152
  mistrust, 14, 16
  multi-, 49
  post-, 124
  preferences, 133
  profidling, 166
  quotas, 162
  segregation, 5, 43, 70, 73, 139, 140
  spokespersons, 128
  stratification, 88
  trust, 3, 52
Racine Correctional Institution, 24
racism, 6, 29, 53, 60, 70–71, 124
  black-white, 72
  legacies of, 60
  structural, 71
  subjective, 71
  white, 177
Regents of University of California v. Bakke, 127–128, 161–162
residence hall
  peace-and-justice, 63
  racially themed, 63
role reversal, 44, 45, 48, 50
  pods, 47
Rutgers University, Newark, 7, 83–88

SAT, 7, 8, 10, 57, 84, 85, 100, 102, 104, 106, 111, 131, 133, 134, 137, 138, 154
  as proxy for SES, 137
Scalia, Antonin, 5, 6, 166
scholarship, 15, 30
  and activism, 30–31
Scholarship in Action, 35–36
Science, 113
science, technology, engineering, and math (STEM) fields, 56, 57, 62
segregation, 94, 148, 150
  academic, 174
  school, 149
  residential, 19, 70, 71, 149, 150, 152, 154, 160, 162
selective colleges, 100–103, 105, 110–111, 114, 131, 132, 134, 137–141, 160, 162
The Shape of the River, 105, 125, 131, 141
Smith, Matthew A., 11, 109, 242
social
  capital, 77, 78, 80
  change, 12, 65, 185
  crisis in Detroit, 72
  diversity, *see under* diversity
  identity, -ies, 23, 29, 195
  interaction, 57
  justice, 6, 10, 52, 58, 60–63, 65, 92, 104, 185, 189, 194
    programs, 59
    role of education in, 58
  mobility, 7, 9, 70, 71, 74
  needs, 7
  opportunity, 71
  resource, 6
  role, 8
  safety net, 9
  status, 8
  transformation, 185, 192
  trust, 3, 17
social identity threat, 195–206
  literature, 196
"social mistrust," 2–4

socially
  embodied beings, 15
  marginalized, 7
  produced strengths, 15
  underprivileged, 3
socioeconomic
  advantages, 110
  background, 163
  deprivation, 11
  disadvantage, 99, 104, 161, 168
  profile, 152
  status (SES), 11, 12, 69, 101, 109, 114, 127, 135, 150, 156
    applicants, 69
    and early childhood, 113
    high- students, 136, 137, 140, 146
    low- students, 73, 102
    and race, 69
solo status, 198–206
Spencer Foundation, 11, 109
standardized test, 11
  scores, 131, 133, 137, 163
  see also ACT scores, Graduate Record Examinations (GRE), SAT
Stanford University, 1, 55, 57–58, 109, 114, 174, 185, 192, 242
  Medical Youth Science Program (SMYSP), 57–58, 66
State University of New York (SUNY), 34, 93
Steele, Claude, 1, 14, 16, 17, 27, 56, 196
stereotype(s), 14, 15, 182–184, 186–187, 198–199, 201–202, 204, 205, 209
  cultural, 182
  images, 14
  negative, 194–196, 198, 200
  racial, 2
    negative, 1, 2
    positive, 2
  social, 3
    negative, 16
"struggler," 198

stereotype(s)—*Continued*
  threat, 2, 16, 195–198, 201, 202
    alleviating, 197, 201
stereotyping
  ethnic, 87
stratification, 99
streetcar university, 84
struggler language, 206
struggler phenomenon, 182, 184–206
  and affiliation, 192–194, 207
  and disidentification, 194–195
  and mentors, 188–192, 193, 207, 209–210
  research on, 195–205
  strategies to address, 206–210
student affairs
  office(s) of, 53
students of color, 12, 15, 51, 136, 160, 164
"successful academics," 15, 182–184, 192, 194, 195, 201, 205, 208
Sugrue, Tom, 71
Sullivan, Teresa A., 11, 13, 147, 242–243
supply-side block, 100
Supreme Court, 4, 19, 41, 86, 102–103, 105, 115, 126–129, 151, 159–161, 163
Syracuse University, 4, 19, 24, 27–29, 32–34, 36, 55, 61–62, 115–116, 192

teaching, 15, 16, 77
Team Against Bias, 29
Texas Top Ten Percent law, 12, 13, 138, 147, 154–155, 160
  controversy around, 154–155
Thompson, Dennis, 113
thumb on the admissions scale, 10, 103, 104
Tilghman, Shirley, 181–182, 183, 201–202, 207, 210
Tobin, Eugene M., 10, 44, 58, 59, 62, 97, 114, 243
tracking, 11, 134–135
tribalism, 6, 43

trust, 17, 23

"undermatch," 108
underrepresented, 15
  groups, 52, 57, 59, 159–160
  minorities, 101–102
university
  mission, 209
  research, 6, 53, 55, 56
  scholarly mission, 65
University of Arizona, 13, 181, 185–186, 190
University of Buffalo, SUNY, 34
University of California
  at Davis, 128
  System, 153, 164
University of Denver, 11, 123
University of Illinois at Urbana-Champaign, 19, 21
  controversy at, 21–22
University of Michigan, 4, 12, 19, 23, 27, 32, 98, 102–105, 129, 147, 155, 159–165, 168
  at Ann Arbor, 54
  -Dearborn, 7, 69, 70, 73–77
  Law School, 5, 41, 45, 128, 159
University of Oregon, 6, 51, 55, 192
University of Texas at Austin, 154, 165
University of the District of Columbia, 83
University of Virginia, 11, 97, 147
Urban Studies department, 83
U.S. News and World Report, 115
  ranking system, 7, 85, 86

Wellesley College, 169
white flight, 32
Wilgoren, Jodi, 24
workers
  immigrant, 8
  urban, 8

Yale University, 103, 109, 114, 160, 174, 242
Young Women's Christian Association (YWCA), 95